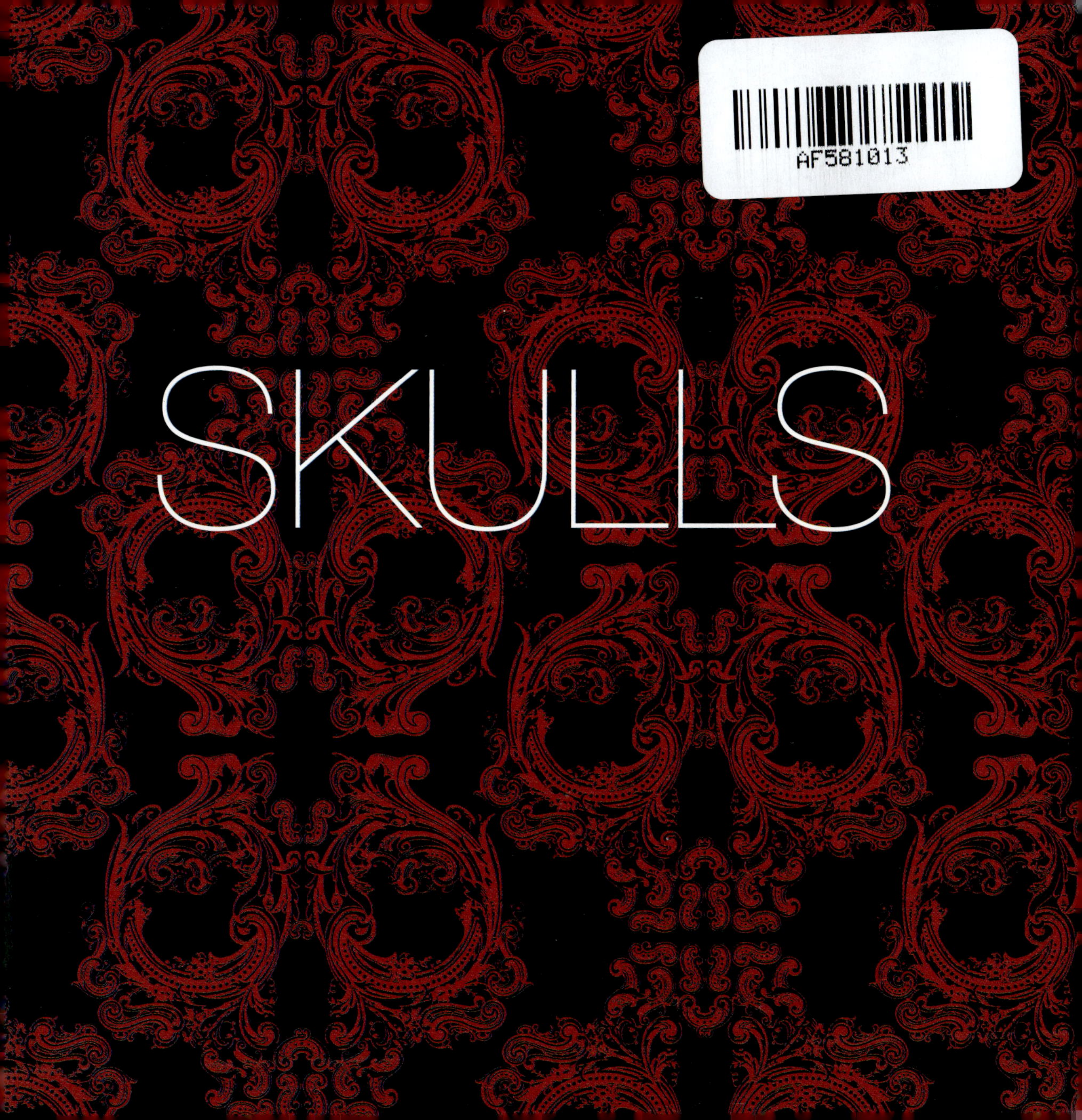

SKULLS

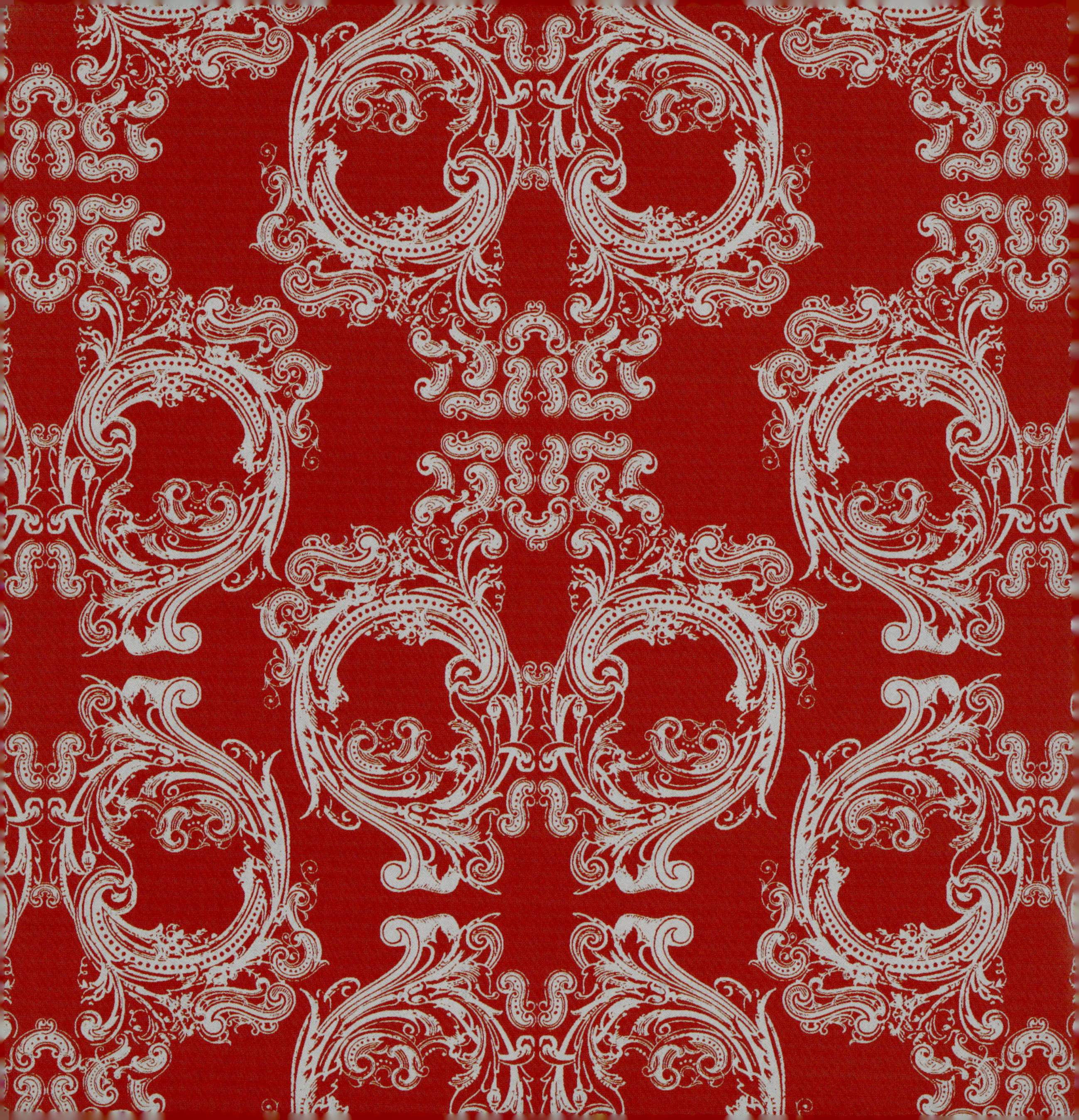

SKULLS

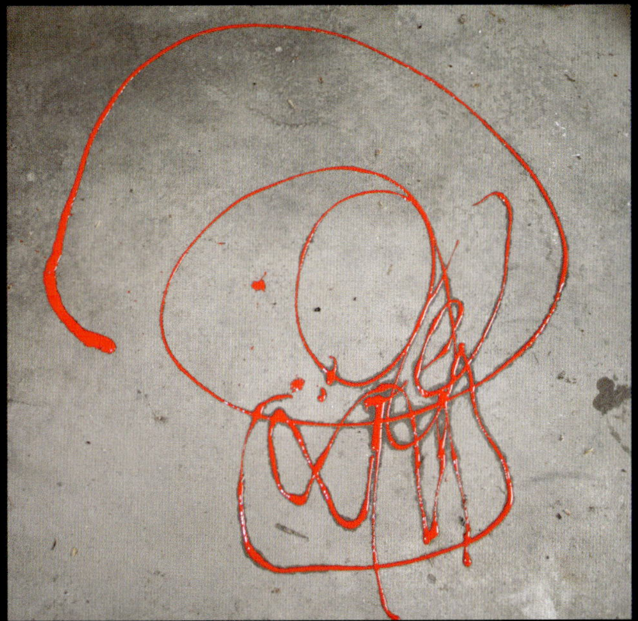

Noah Scalin

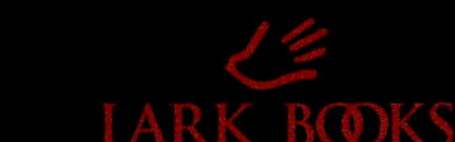

A Division of Sterling Publishing Co., Inc.
New York / London

Dedication:
This book is dedicated to the fans of the Skull-A-Day website. Turning my whim into something tangible would not have been possible without your enthusiasm and encouragement.

SENIOR EDITOR:
Deborah Morgenthal

BOOK DESIGN:
susanmcbridedesign.com

PHOTOGRAPHER:
Noah Scalin

COVER:
Chris Bryant
Noah Scalin

Library of Congress Cataloging-in-Publication Data

Scalin, Noah.
Skulls / Noah Scalin.
p. cm.
Includes index.
ISBN 978-1-60059-375-8 (pb-trade pbk. : alk. paper)
1. Handicraft. 2. Death in art. 3. Skull. I. Title.
TT157.S29 2008
745.5--dc22
2008009963

10 9 8 7 6 5 4 3 2

Published by Lark Books, A Division of
Sterling Publishing Co., Inc.
387 Park Avenue South, New York, NY 10016

Distributed in Canada by Sterling Publishing,
c/o Canadian Manda Group, 165 Dufferin Street
Toronto, Ontario, Canada M6K 3H6

Distributed in the United Kingdom by GMC Distribution Services,
Castle Place, 166 High Street, Lewes, East Sussex, England BN7 1XU

Distributed in Australia by Capricorn Link (Australia) Pty Ltd.,
P.O. Box 704, Windsor, NSW 2756 Australia

If you have questions or comments about this book, please contact:
Lark Books
67 Broadway
Asheville, NC 28801
828-253-0467

Manufactured in China

ISBN 13: 978-1-60059-375-8

For information about custom editions, special sales, premium and corporate purchases, please contact Sterling Special Sales Department at (800) 805-5489 or specialsales@sterlingpub.com.

contents

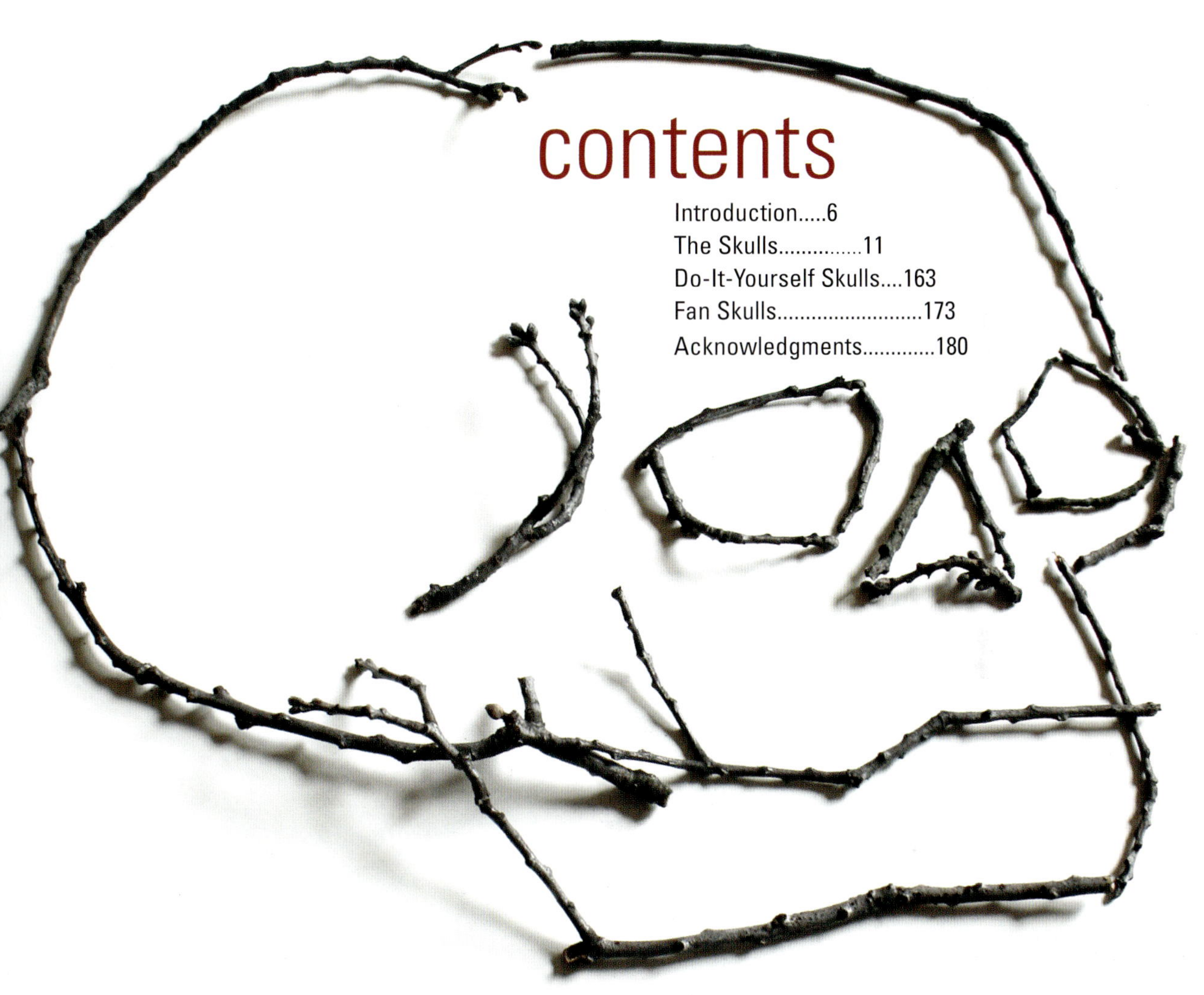

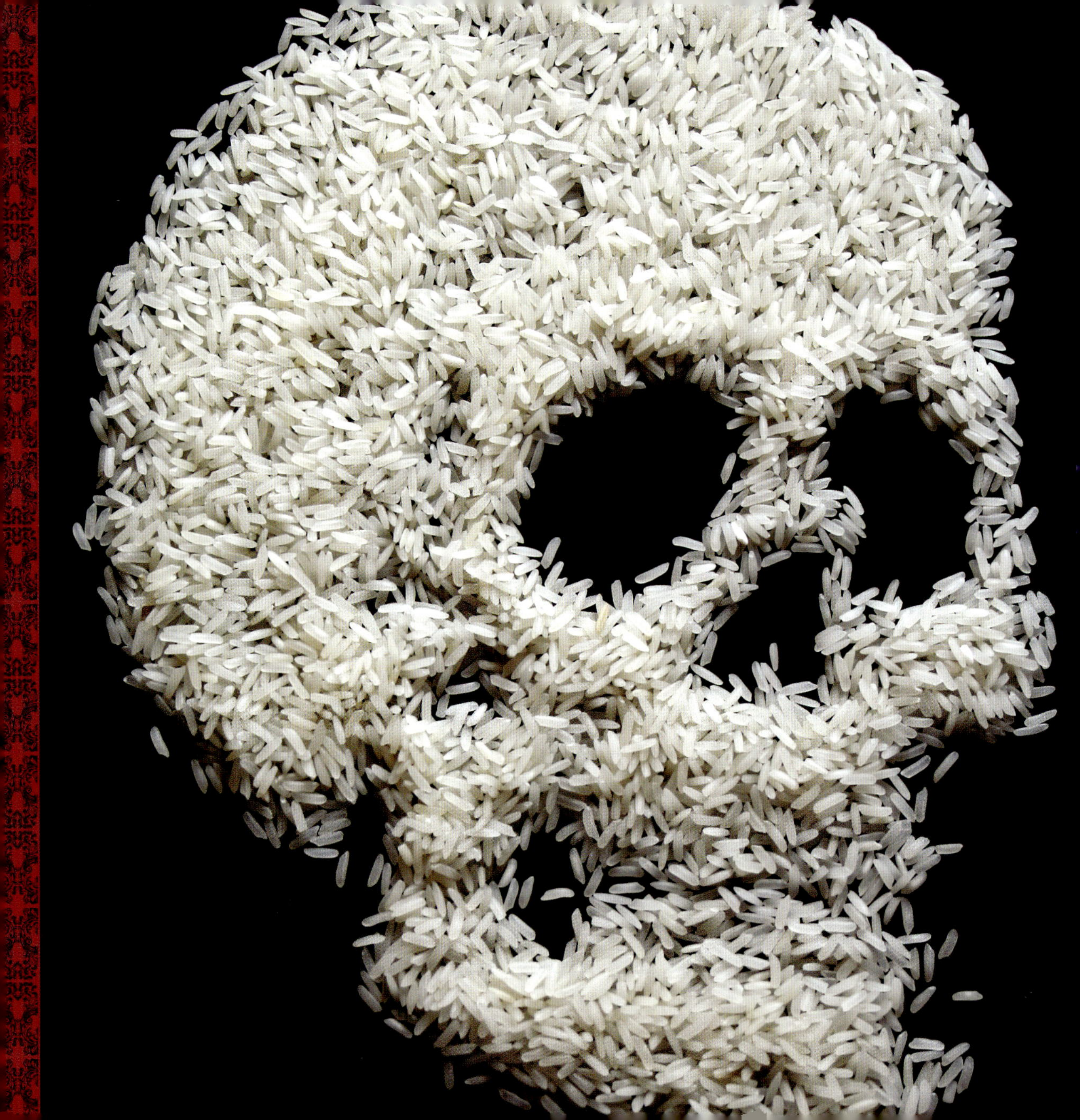

introduction

On June 4th, 2007 I cut a skull out of orange paper and posted it online with the note, "I'm making a skull image every day for a year." This was definitely one of those "easier said than done" situations. Had I realized what it really means to commit yourself to creating an original work of art every day for a year, spending anywhere from two to eight hours a day on it, I might have thought twice about doing it. Then again, had I known I would get dozens of skulls e-mailed to me every week by strangers who were inspired by my project; be mailed original works of art based on my skulls from as far away as Australia; and ultimately asked to create a book to share my work with an even wider audience, I would most likely have done it anyway!

Perhaps surprisingly, there really was no grand impetus for the project. The idea just popped into my head: "Why not make a skull a day for a year?" And since I didn't have a good reason not to…I did.

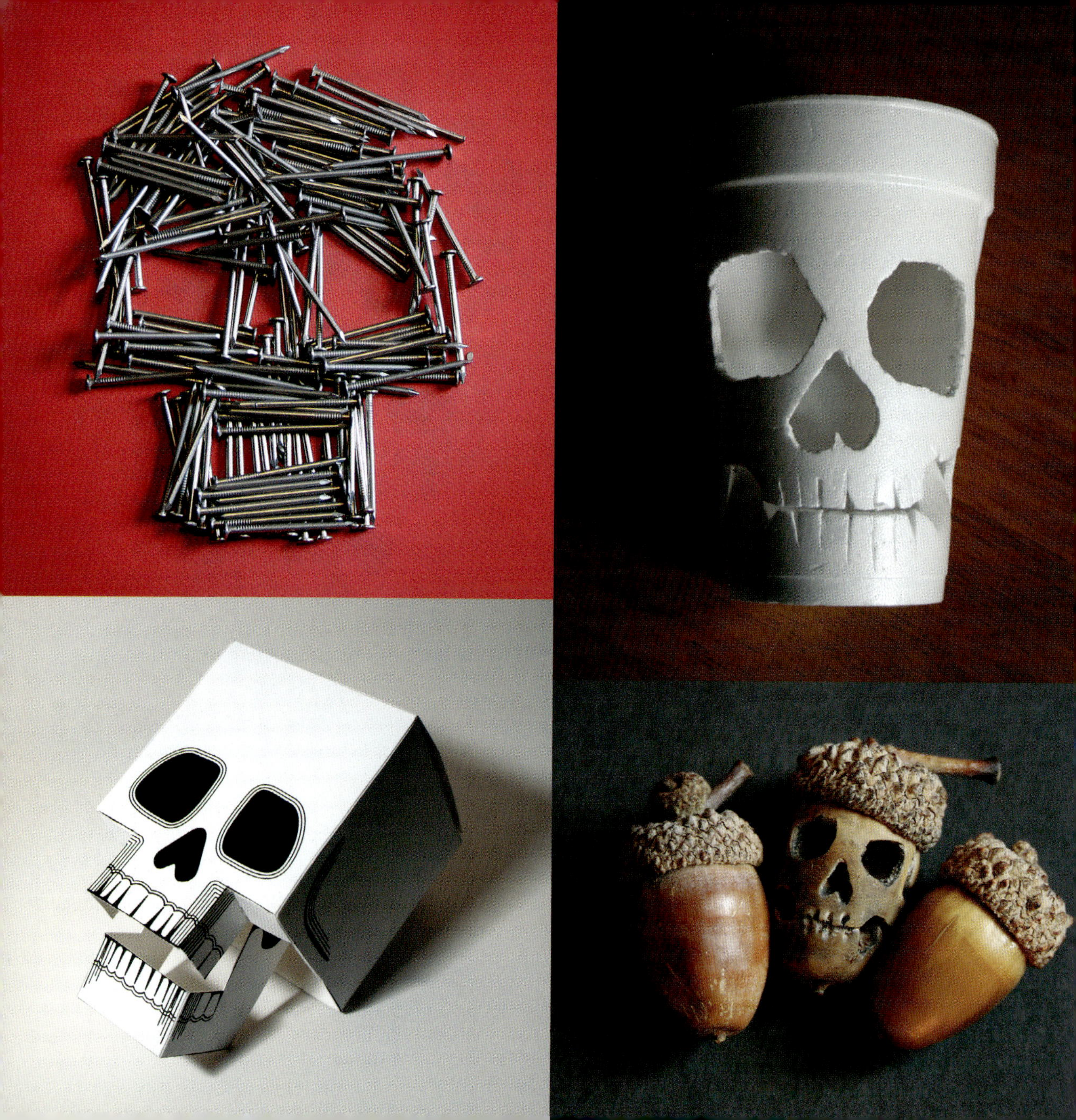

So why skulls? I've always found them fascinating. Heck, I decorate my house with them. I even have several tattoos of them, but it's no morbid fascination. If anything, it's the opposite. The skull has had a respected place in the history of art since the earliest dawning of human creative expression, as a reminder to appreciate the gift of life. It's from this place that my interest springs.

Until relatively recently most people around the world have had a healthy relationship with the reality of death and dying. Mexicans have celebrated *Dia De Los Muertos* (The Day of the Dead) in some form for thousands of years, and the *memento mori* (reminder of death) has been a part of European culture and art since ancient Rome. Indeed, the human-bone decorated Ossuary in Sedlec, in the Czech Republic, and the skeleton-filled Paris Catacombs are still popular tourist destinations. However in the U.S., where death is mostly hidden away in sterile hospitals, the image of the skull has come to represent something morbid, dangerous, or evil, and is only used in popular culture around Halloween.

Of course that has changed some in the last few years as skulls have become part of the mainstream via recent trends in art (Damien Hirst's diamond-crusted skull being the most well known) and fashion (everything from runway garments to Wal-Mart T-shirts). Once the domain of Goths (the subculture, not the ancient East Germanic tribes), skull-clad clothing and jewelry are now nearly ubiquitous. However, little of it carries the weight of Hamlet pondering his deceased friend Yorick while cradling his cranium.

Lacking a societal alternative, my project became my own daily meditation on death. A meditation, mind you, that required a continuous stream of inspiration. And so I mined every nook and cranny of my life for materials to work with. Nothing was safe from death. Every mundane object in my immediate surroundings, from disposable plastic bags to locally grown vegetables, took on deeper significance when transformed into a skull. When nothing was at my fingertips, I scoured art stores, craft stores, and thrift stores for materials to mold and objects to rework. Travel opened up new opportunity for environments to deconstruct—from hotel rooms to construction areas. My favorite experiences, however, involved working with my fellow artists who taught me their skills, like stained-glass making and metal working; and the generosity of friends who allowed me to draw in lipstick in their office bathroom, or rearrange the comic books in their store.

After a year of skulls I can't say that I have truly come to accept the inevitable end of life that we all must experience one day, but I did learn something about living. Every day of that year, I had to have a new experience, find something new in my environment, create something unique, and thus I lived much more engaged in the present than I ever had before. My skulls ultimately succeeded in their role as memento mori by helping me follow the well-known words of the ancient poet Horace, "Carpe diem"… seize the day!

The 150 skulls I chose for this book represent the broad range of materials, techniques, and attitudes I tried out every day. To view the entire year's work, please check out my website, www.skulladay.com. You'll also get to see some of the hundreds of skulls and skull imagery sent to me by people all over the world. A small collection of these is featured at the end of this book. For those of you who want to replicate a few of the skulls I created, you can follow the instructions on pages 164-171 and make four skulls.

I hope you'll be inspired to make your own unique skulls as well.

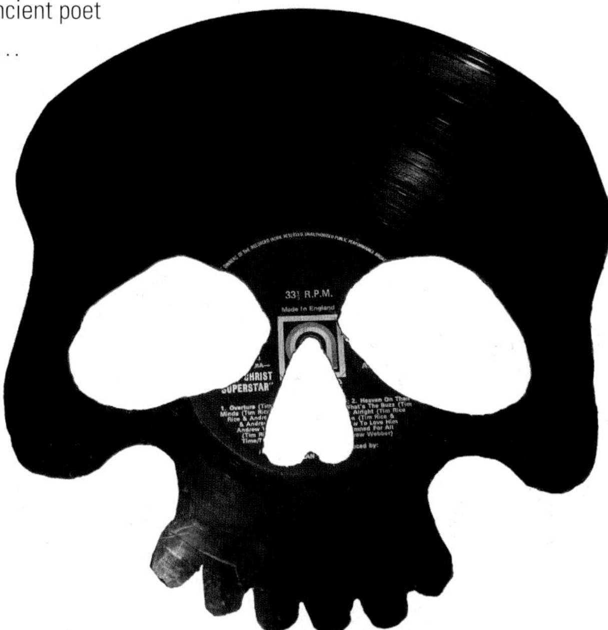

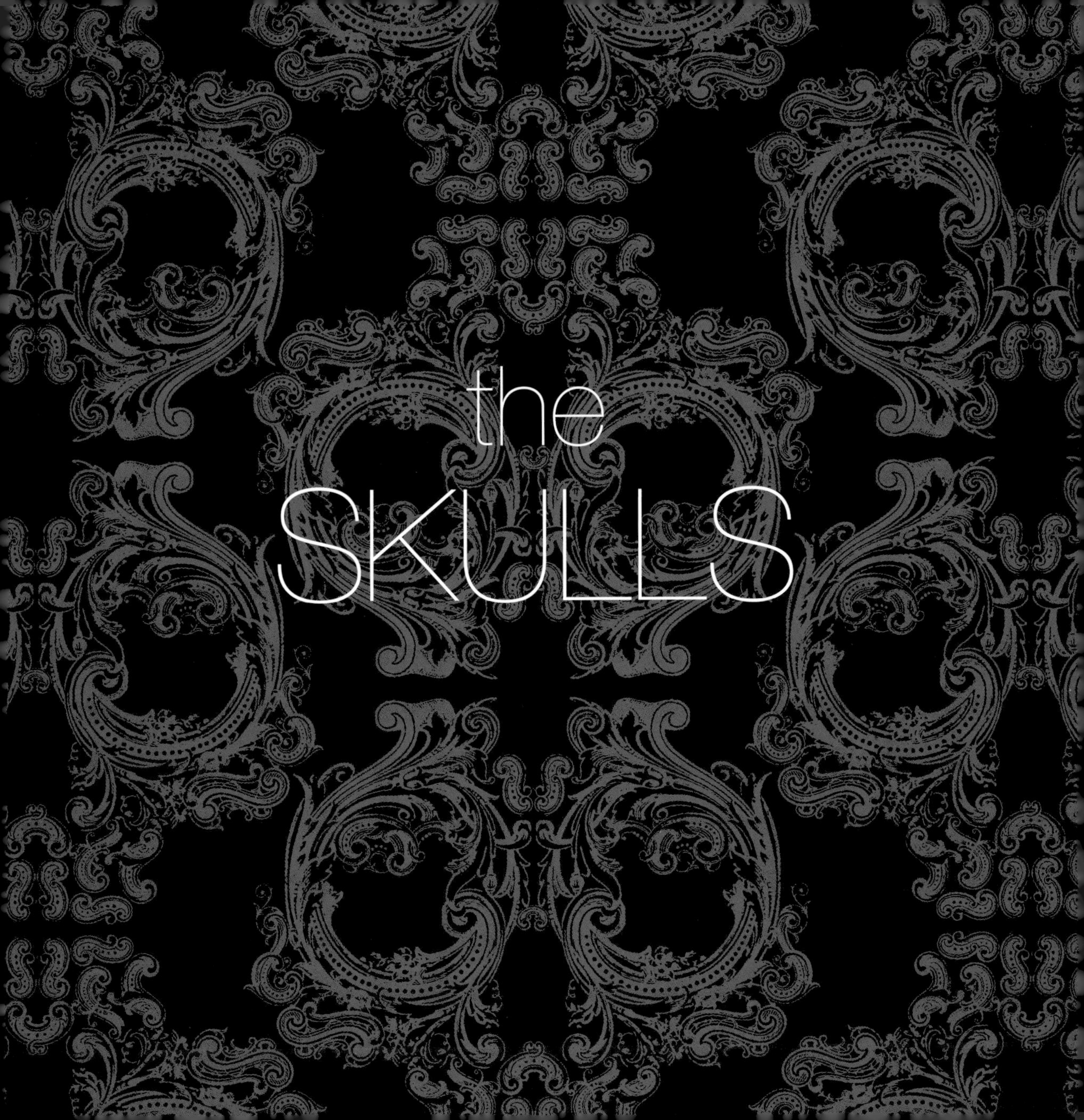
the
SKULLS

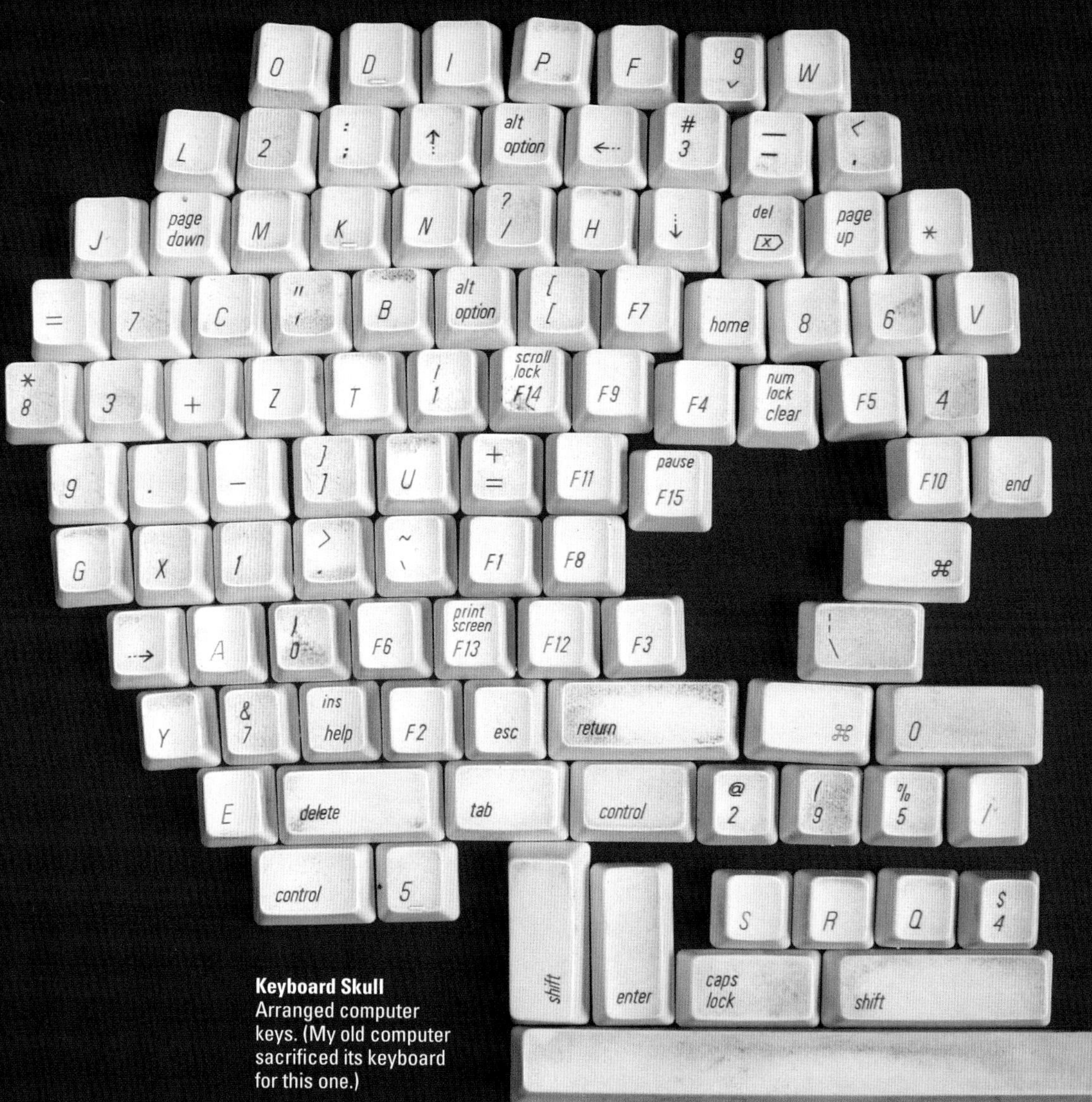

Keyboard Skull
Arranged computer keys. (My old computer sacrificed its keyboard for this one.)

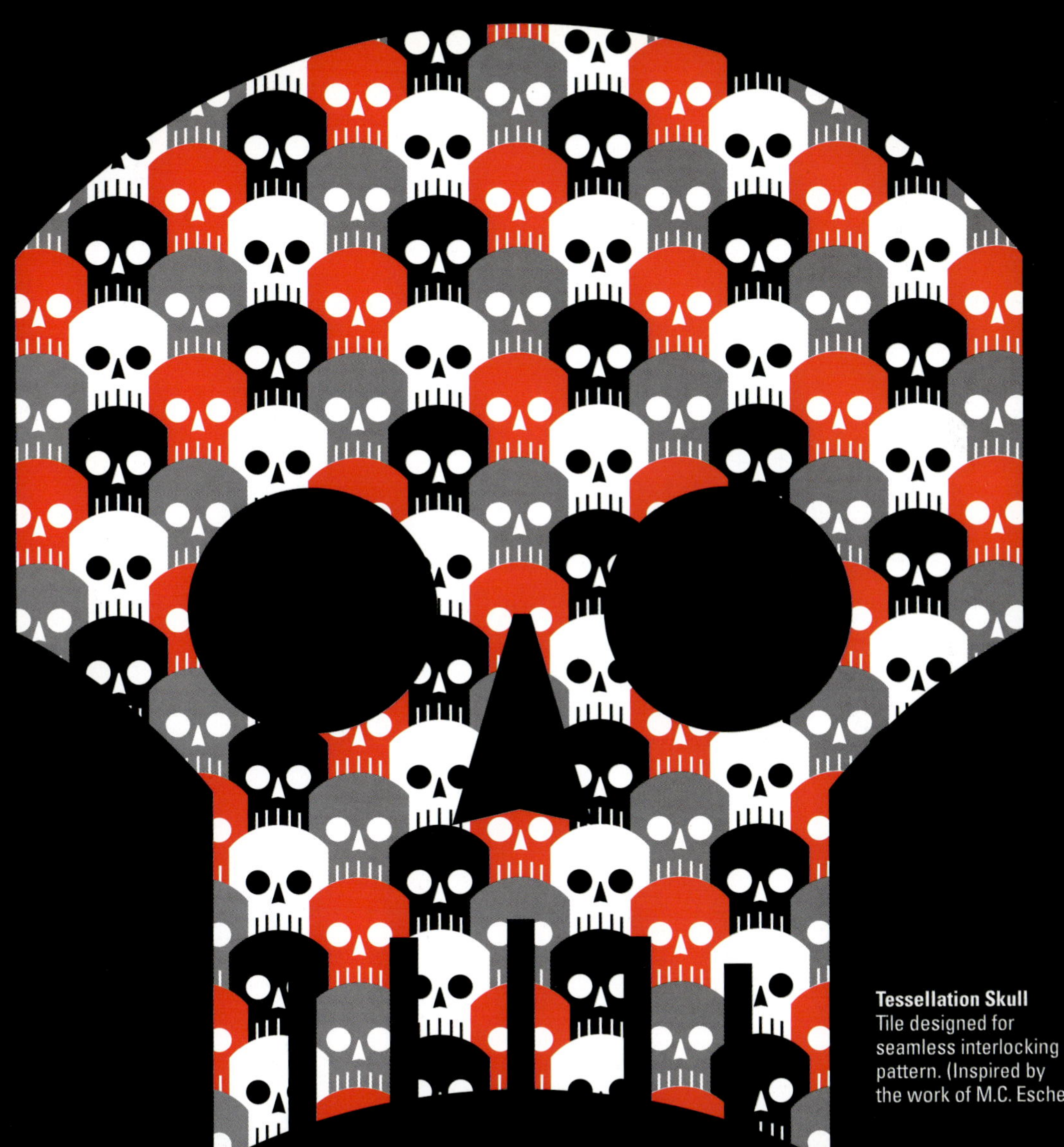

Tessellation Skull
Tile designed for seamless interlocking pattern. (Inspired by the work of M.C. Escher)

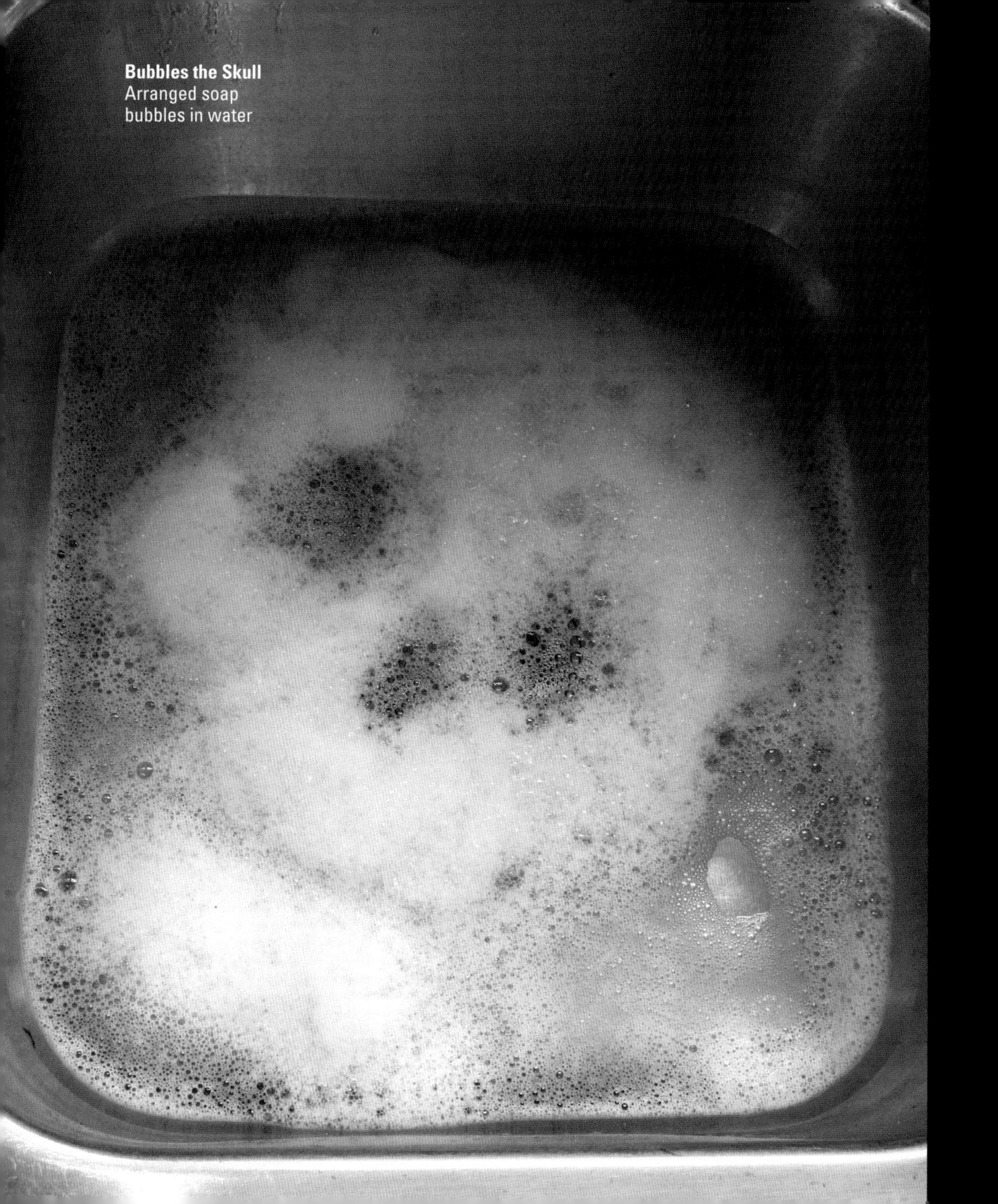

Bubbles the Skull
Arranged soap
bubbles in water

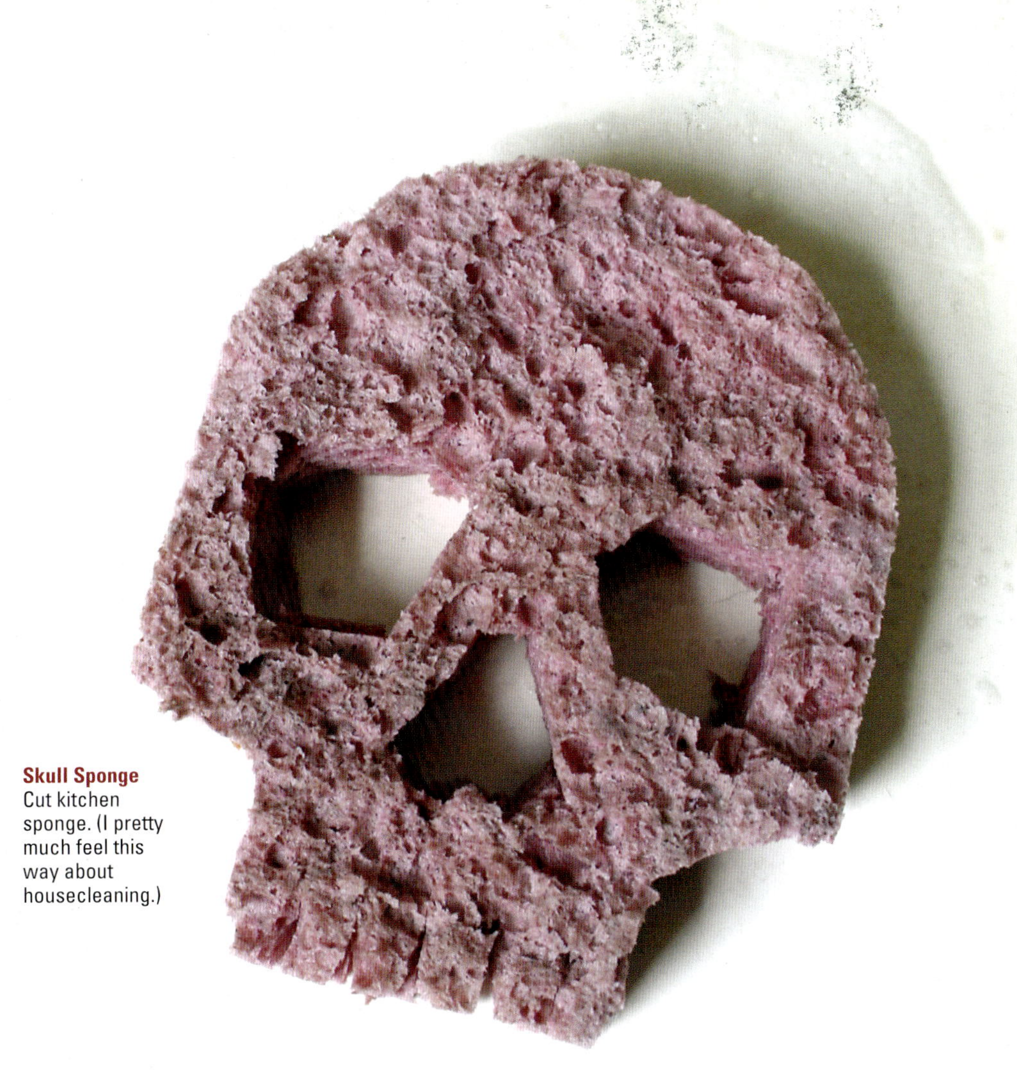

Skull Sponge
Cut kitchen sponge. (I pretty much feel this way about housecleaning.)

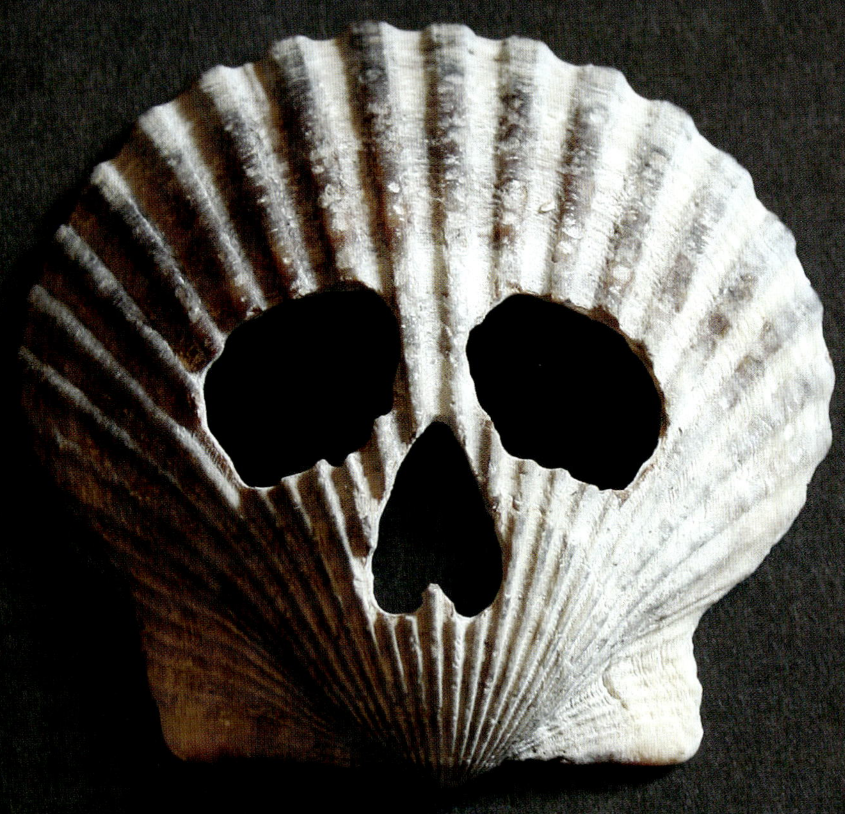

She Sells Seaskulls by the Seashore Cut seashell

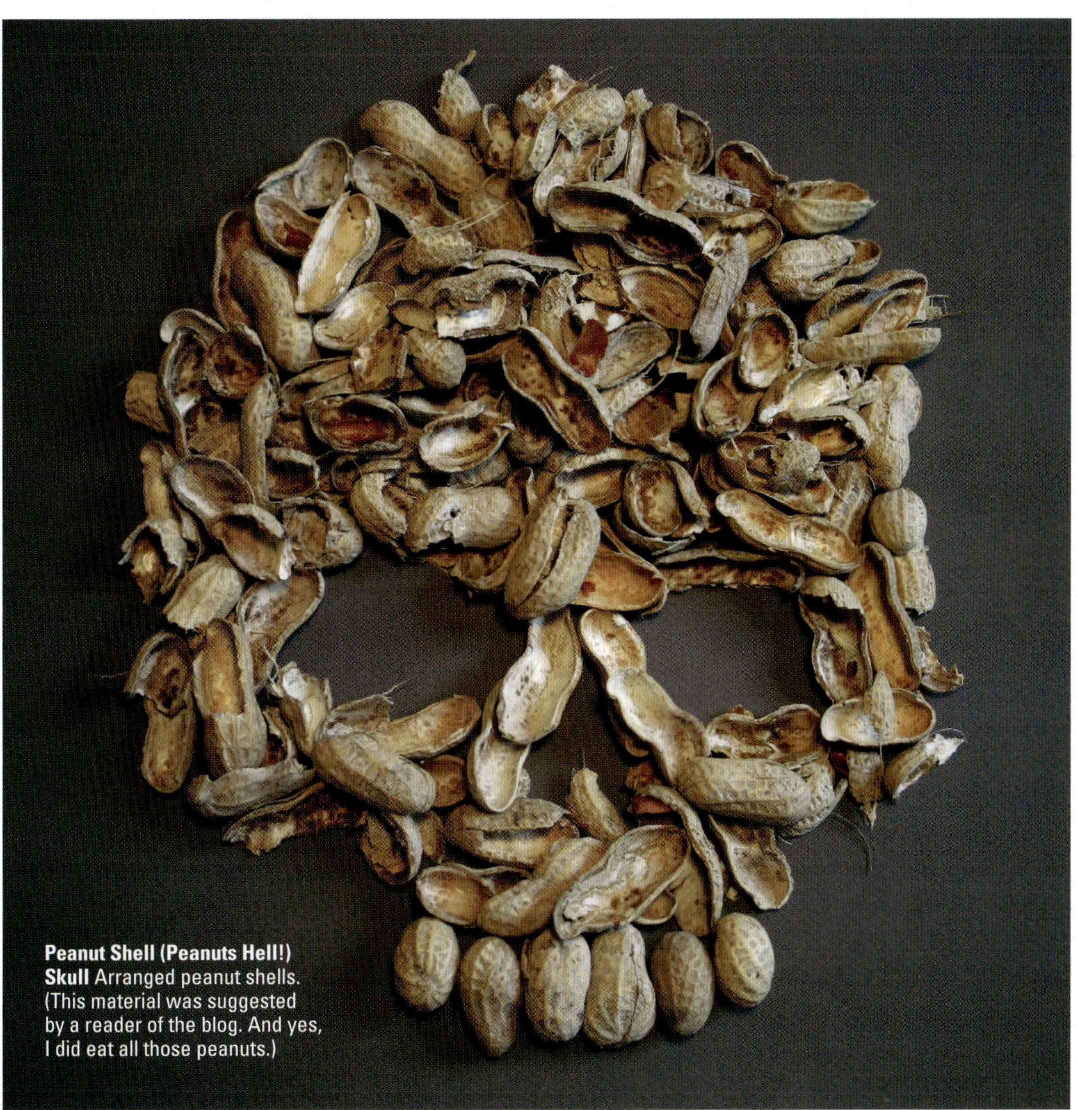

Peanut Shell (Peanuts Hell!) Skull Arranged peanut shells. (This material was suggested by a reader of the blog. And yes, I did eat all those peanuts.)

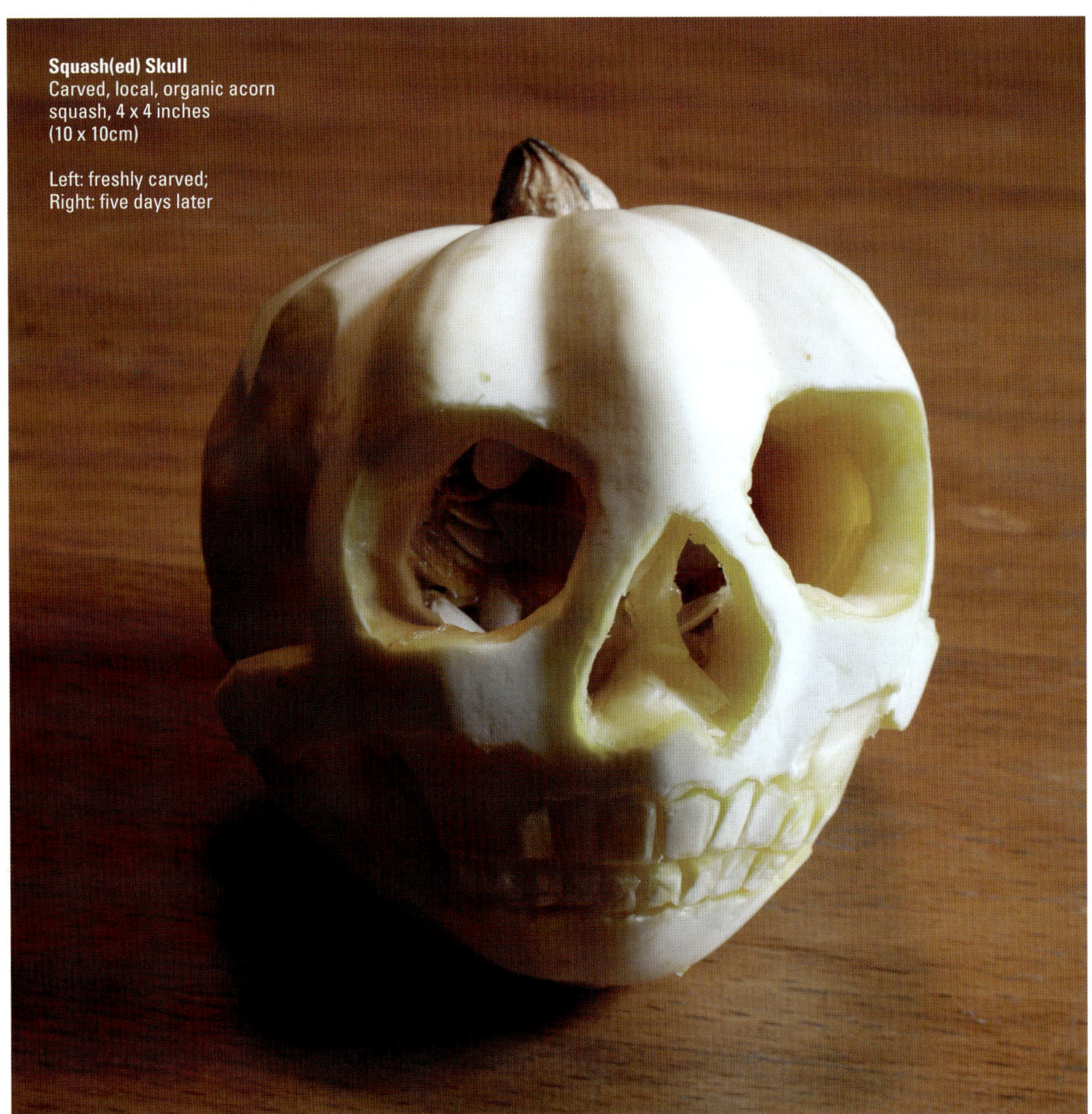

Squash(ed) Skull
Carved, local, organic acorn squash, 4 x 4 inches (10 x 10cm)

Left: freshly carved;
Right: five days later

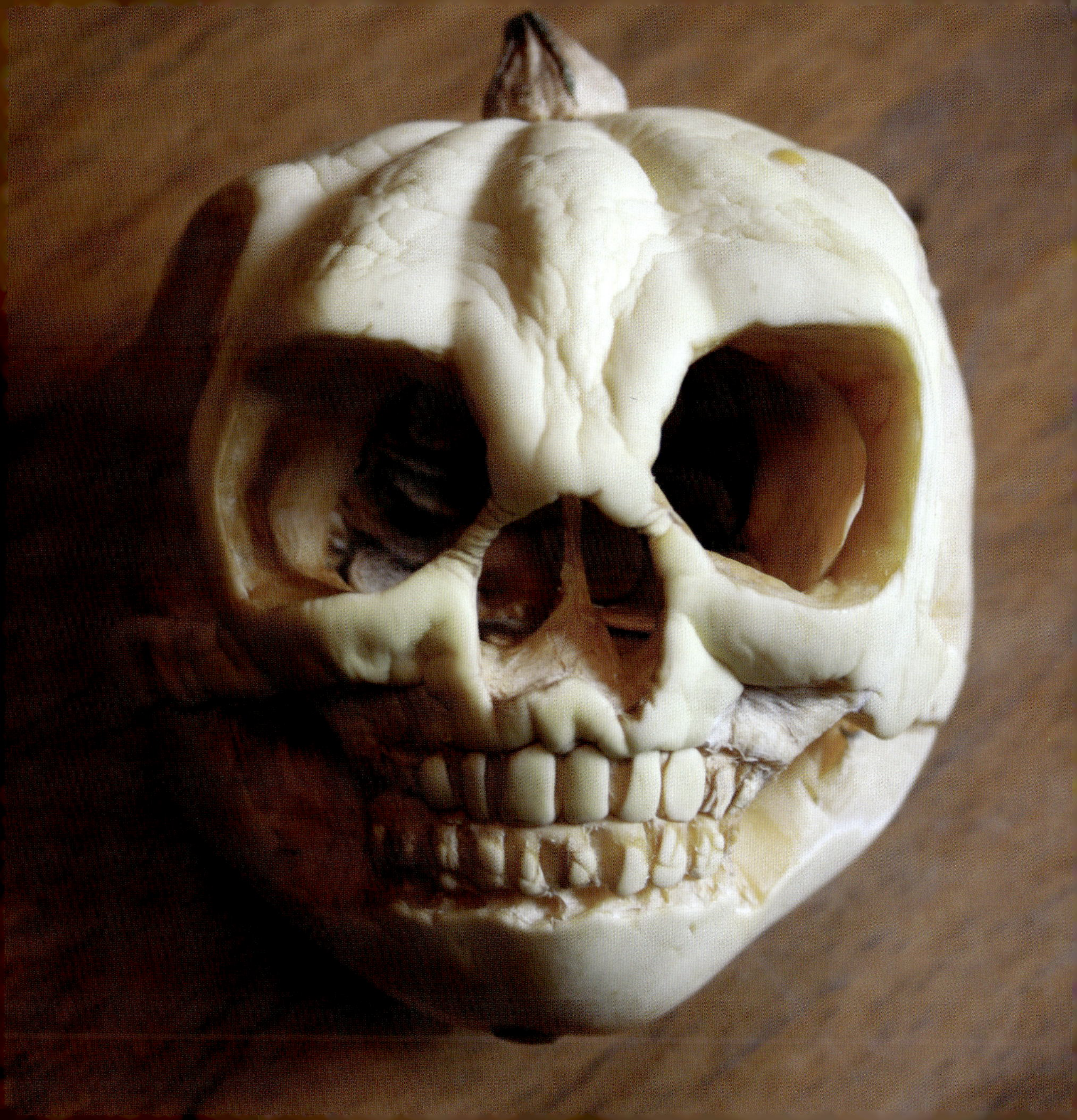

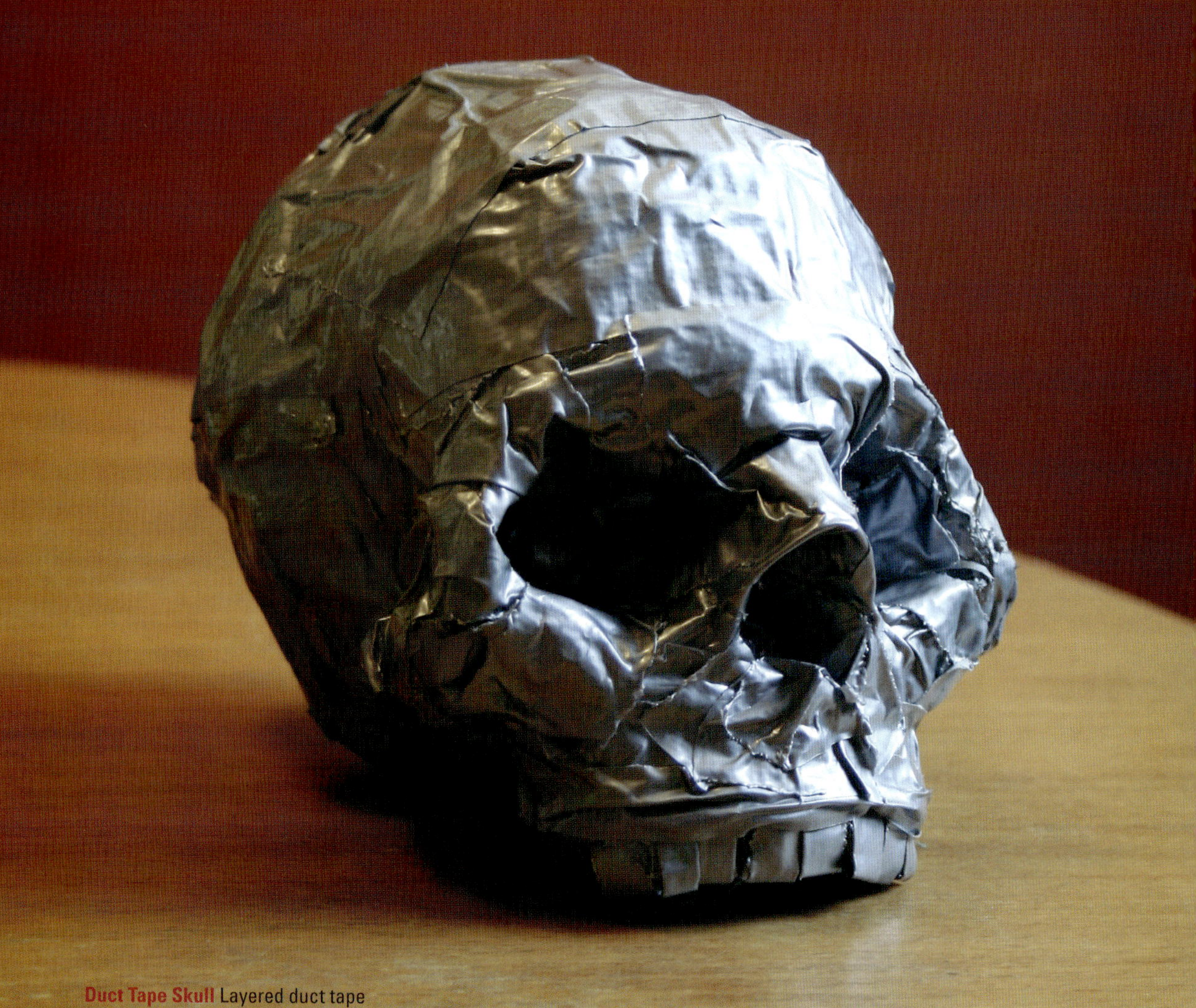

Duct Tape Skull Layered duct tape
5⅛ x 4 x 5½ (13 x 10 x 14cm).
(Is there anything you can't do with
duct tape?)

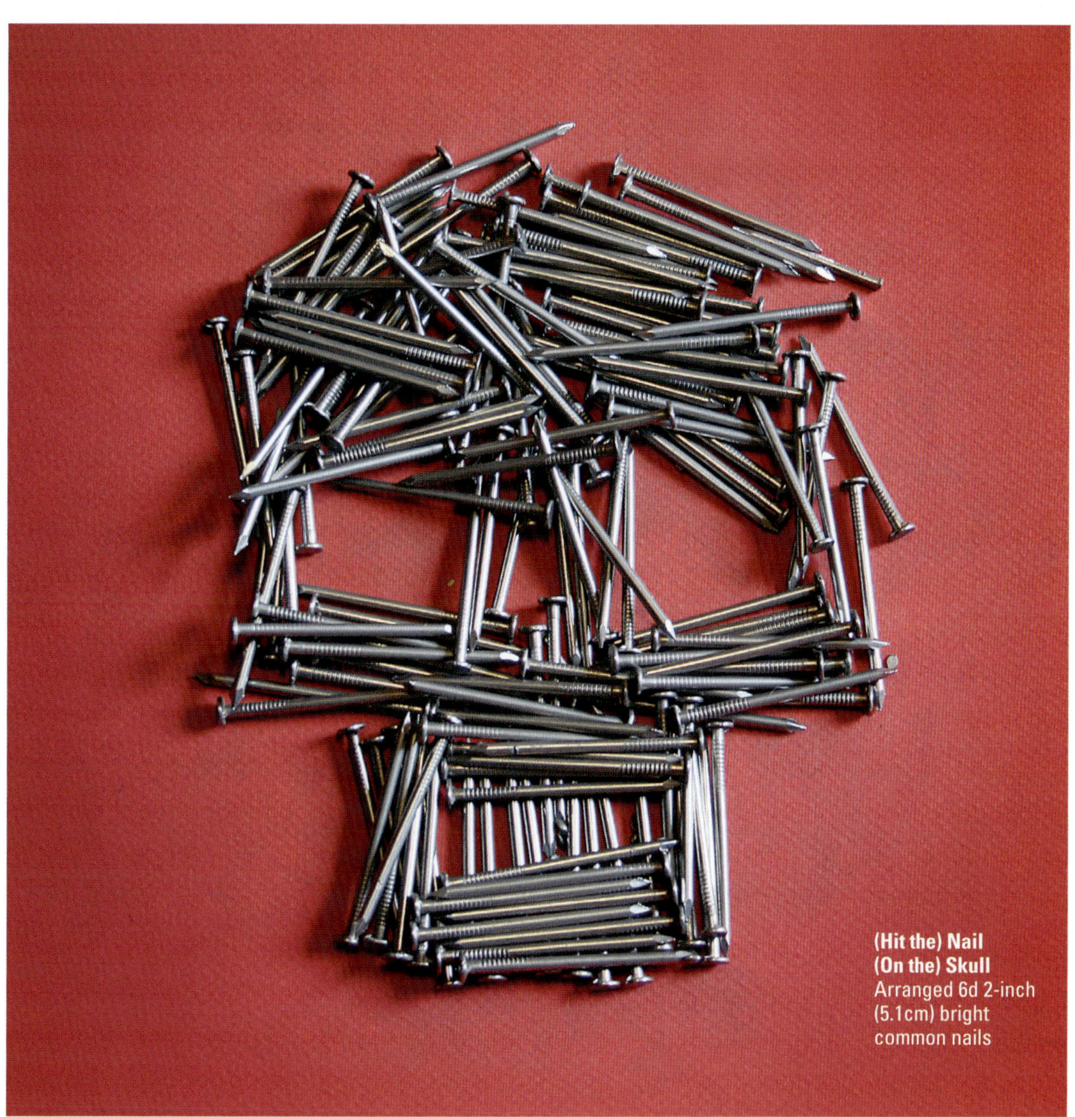

(Hit the) Nail
(On the) Skull
Arranged 6d 2-inch (5.1cm) bright common nails

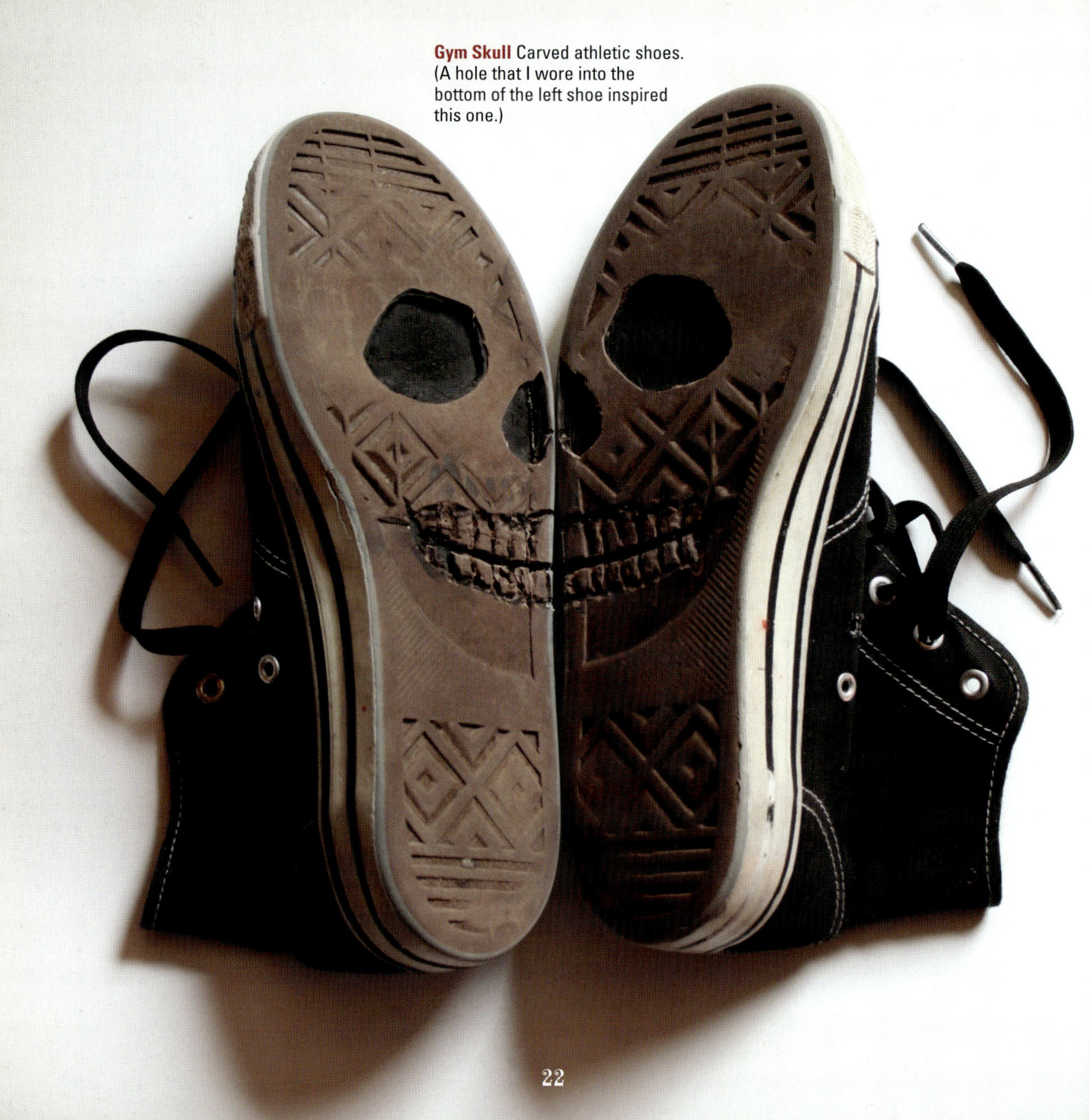

Gym Skull Carved athletic shoes. (A hole that I wore into the bottom of the left shoe inspired this one.)

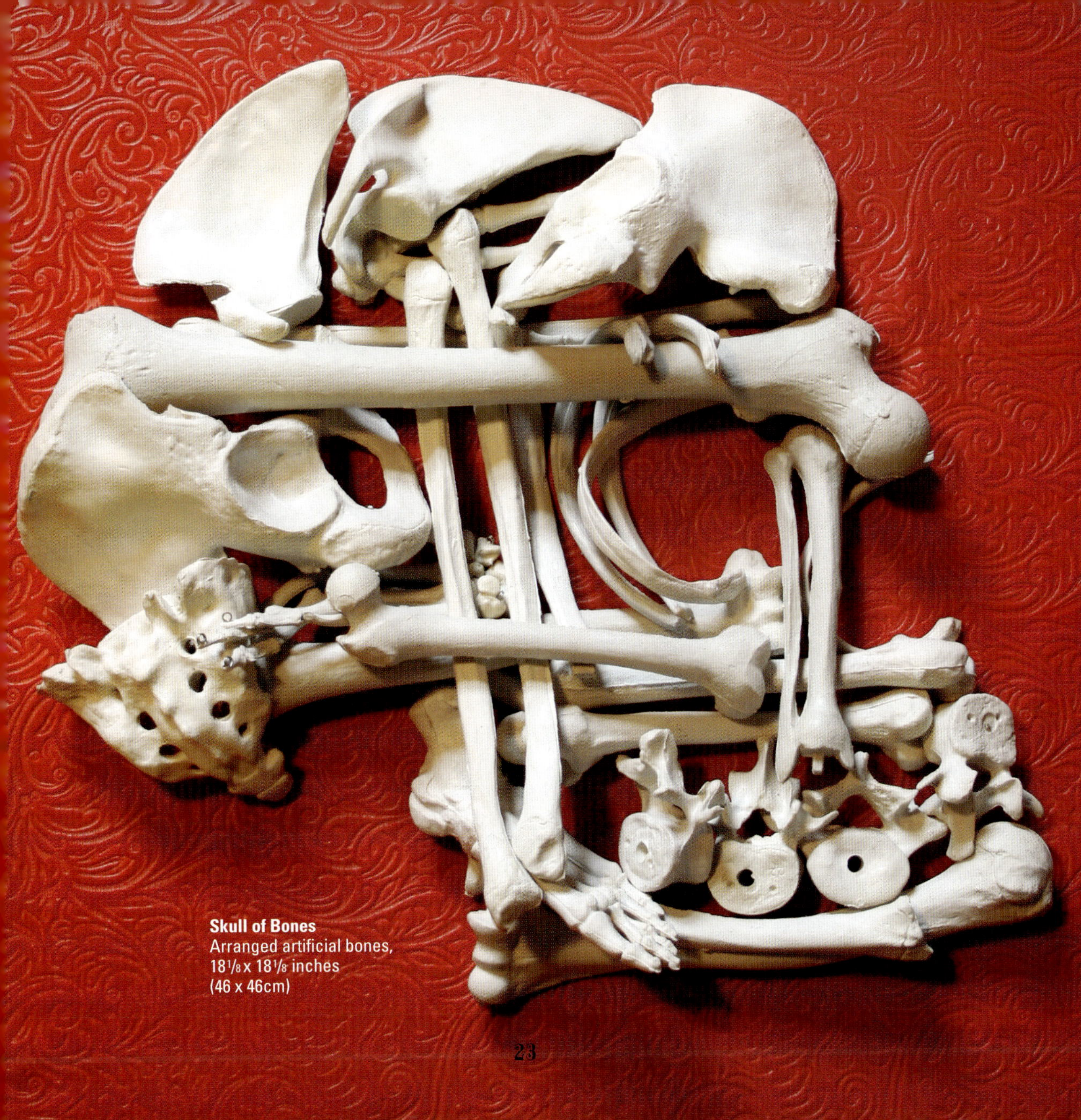

Skull of Bones
Arranged artificial bones,
$18^{1}/_{8}$ x $18^{1}/_{8}$ inches
(46 x 46cm)

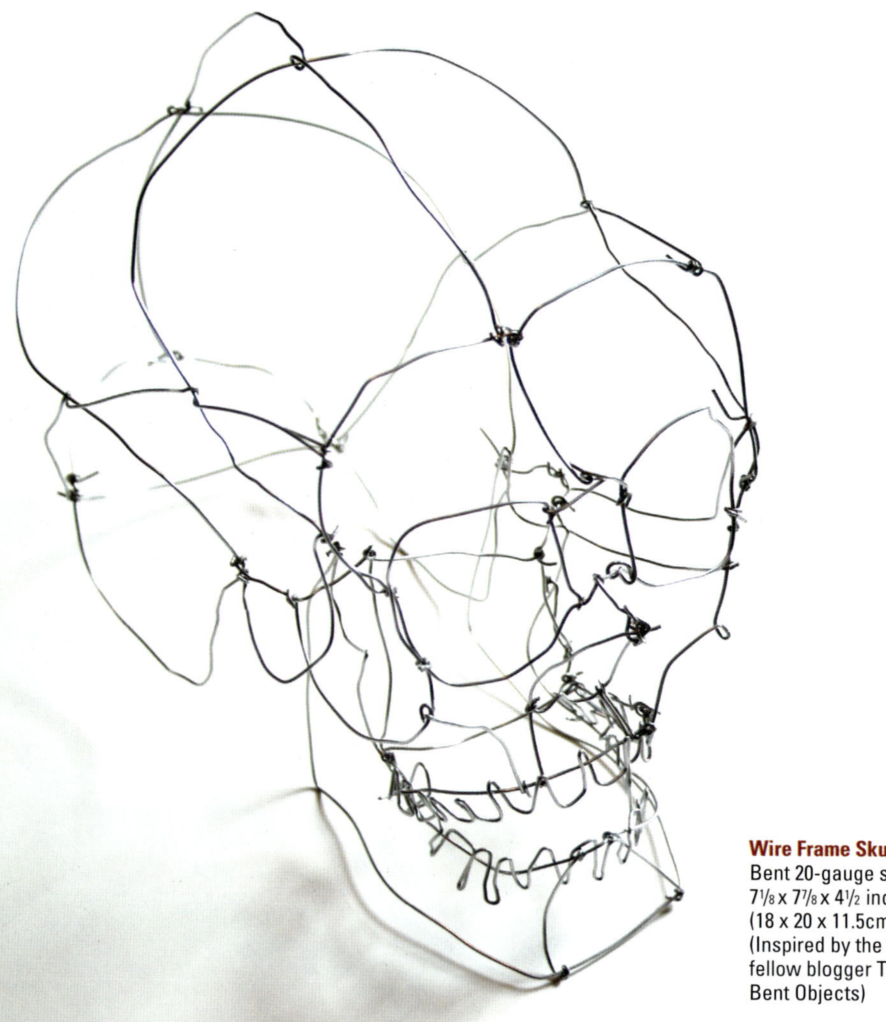

Wire Frame Skull:
Bent 20-gauge steel wire, $7^{1}/_{8}$ x $7^{7}/_{8}$ x $4^{1}/_{2}$ inches (18 x 20 x 11.5cm). (Inspired by the work of fellow blogger Terry at Bent Objects)

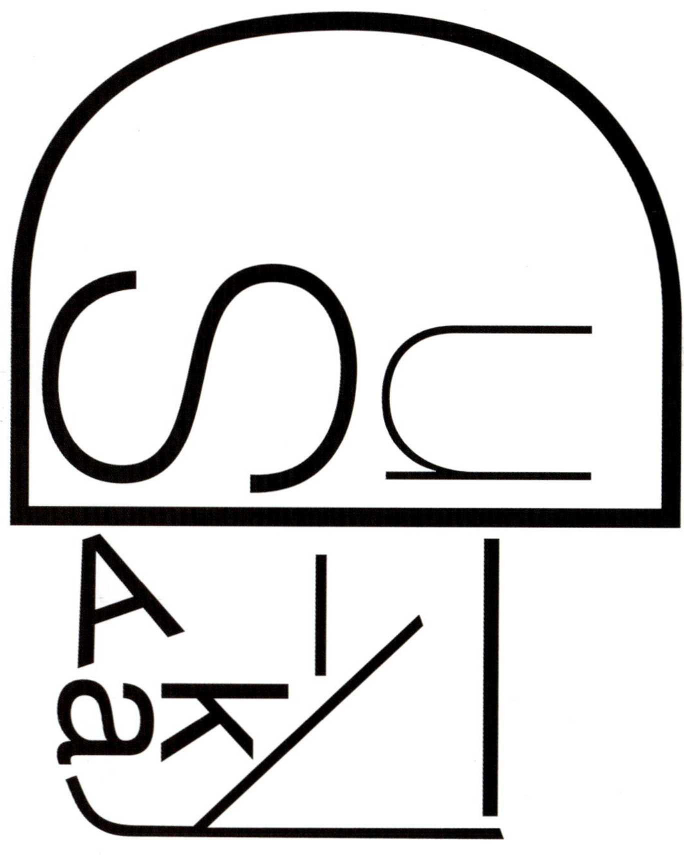

Helvetiskull Arranged Helvetica Neue type. (I limited myself to the letters "Skull A Day.")

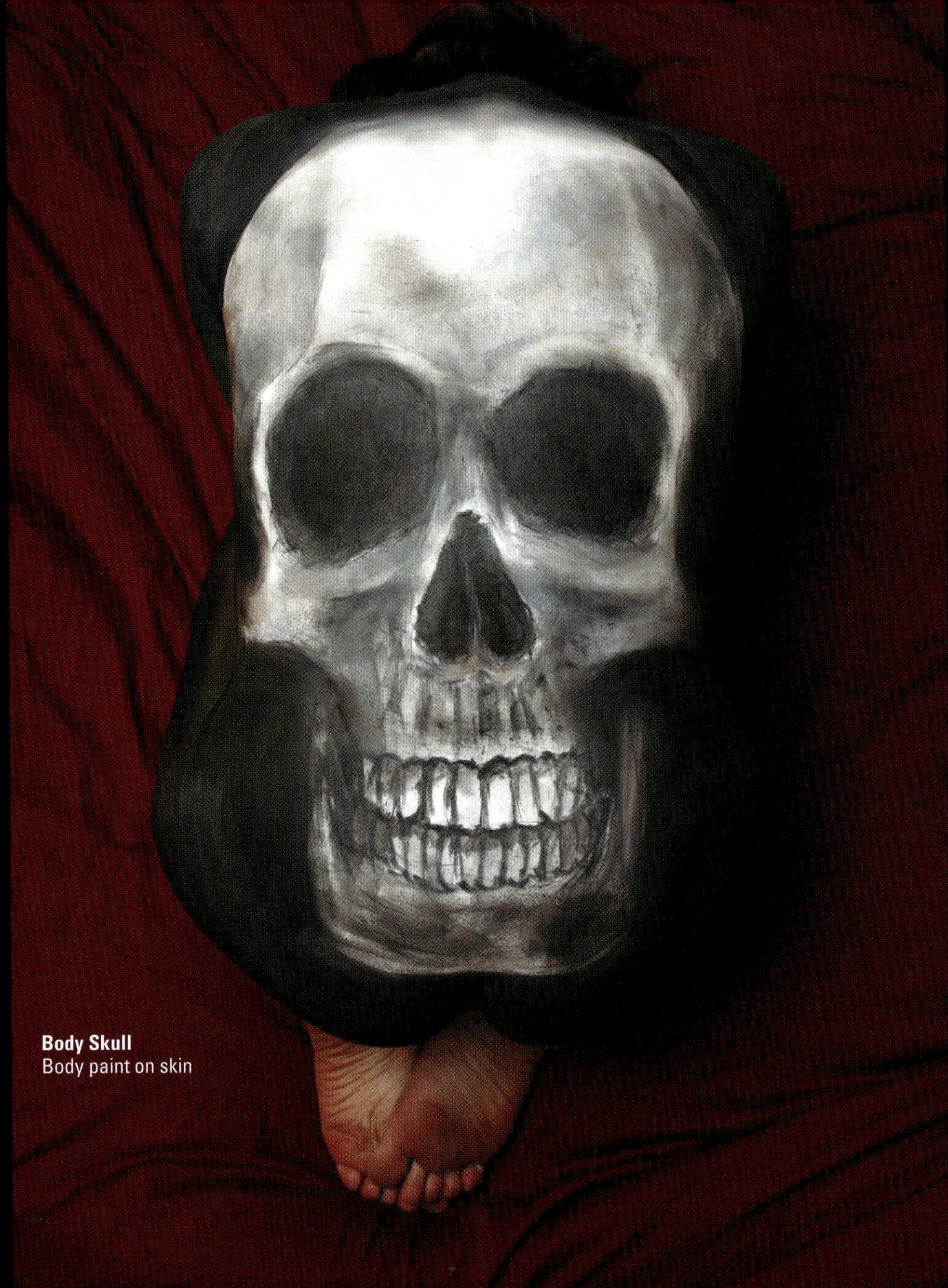

Body Skull
Body paint on skin

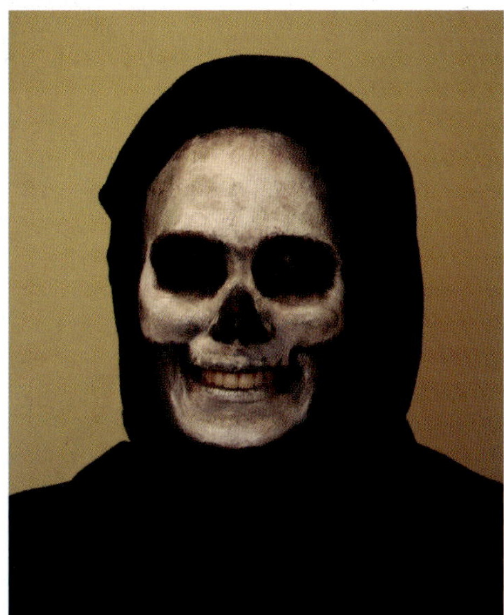

Face Paint Skull Grease paint makeup on skin. (Leftover low-quality makeup from a previous Halloween costume)

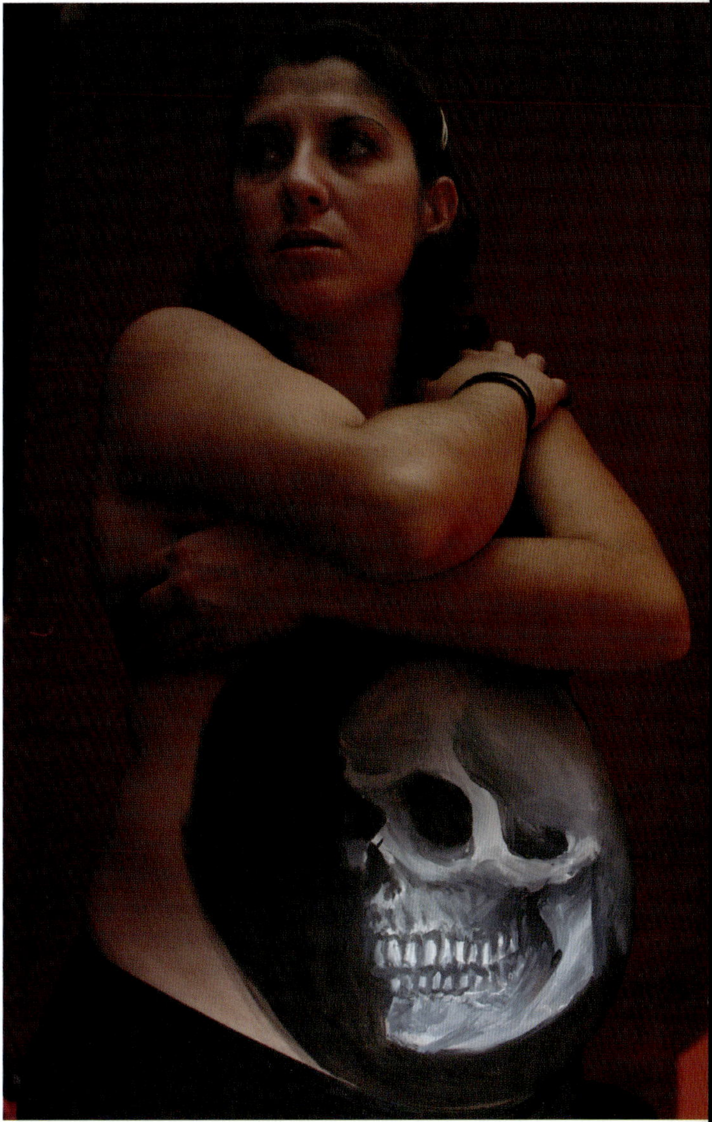

Baby Belly Skull(y) Body paint on skin. (My friend Sara asked me to paint her very pregnant belly for a skull-themed Halloween party I had. Of course I told her it would be a part of my project, to which she replied, "I was hoping you'd say that!")

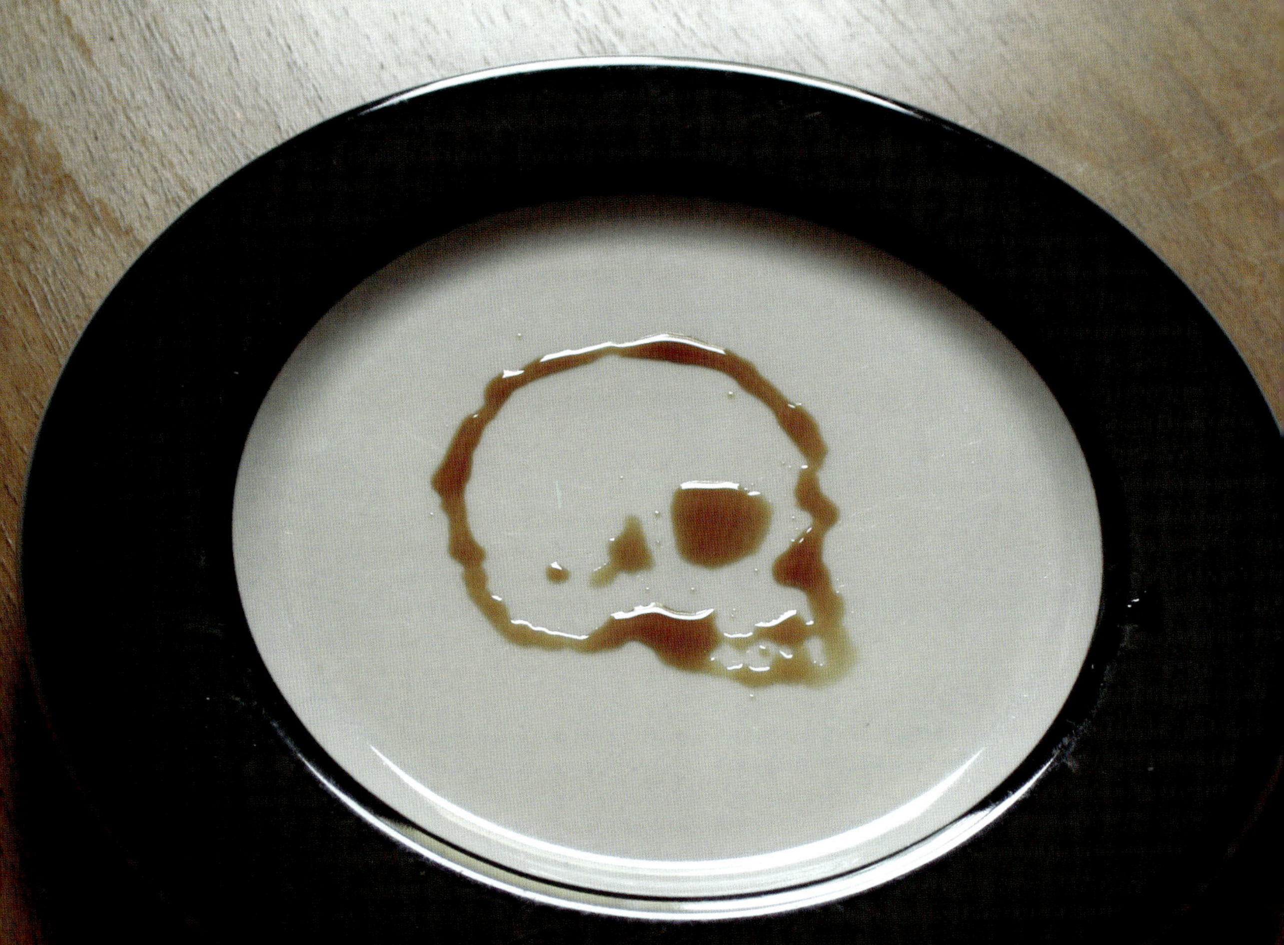

Soy Sauce Skull
Organic soy sauce arranged with a chopstick. (A tricky surface and medium, as it kept changing shape as it slowly moved towards the center of the plate.)

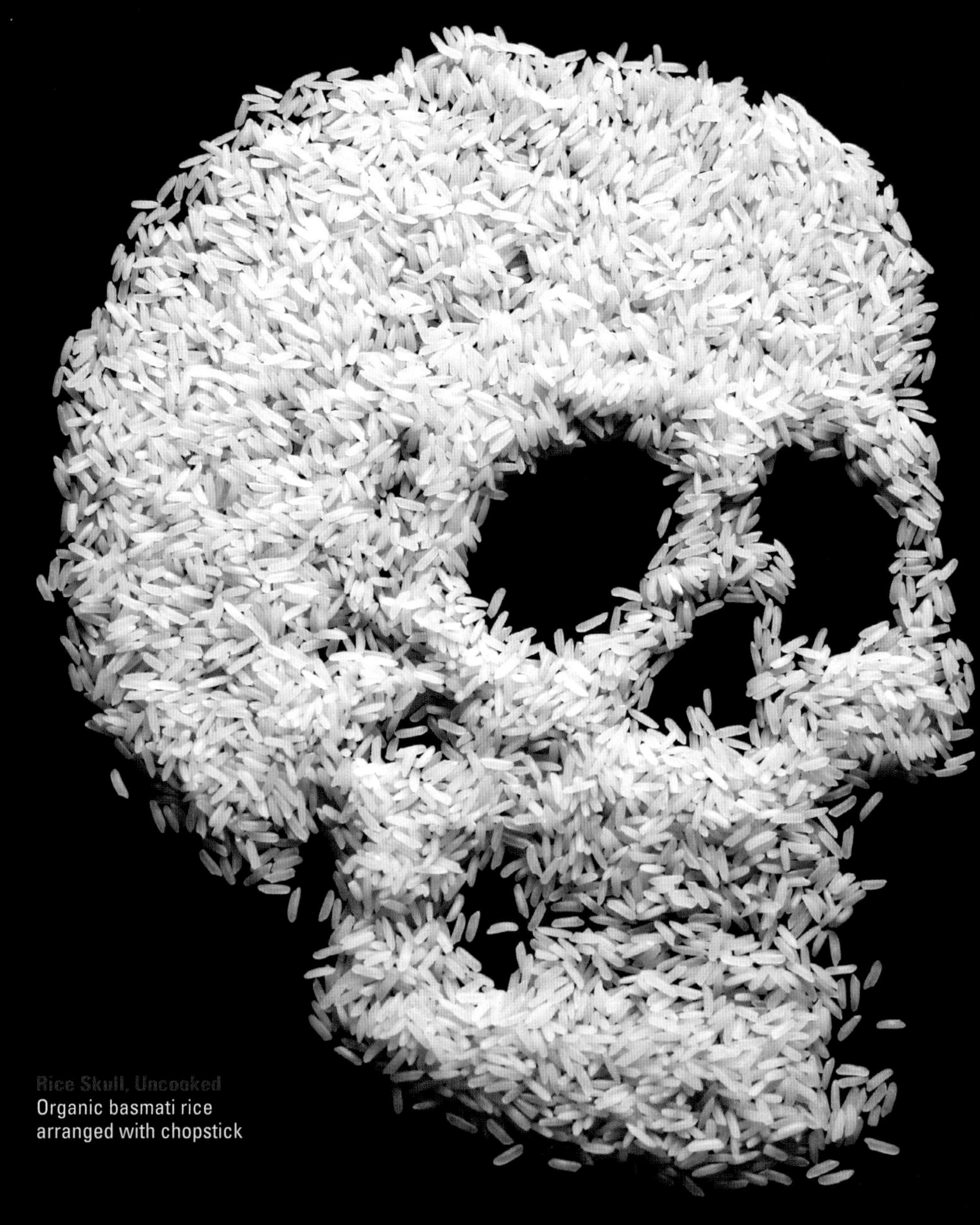

Rice Skull, Uncooked
Organic basmati rice
arranged with chopstick

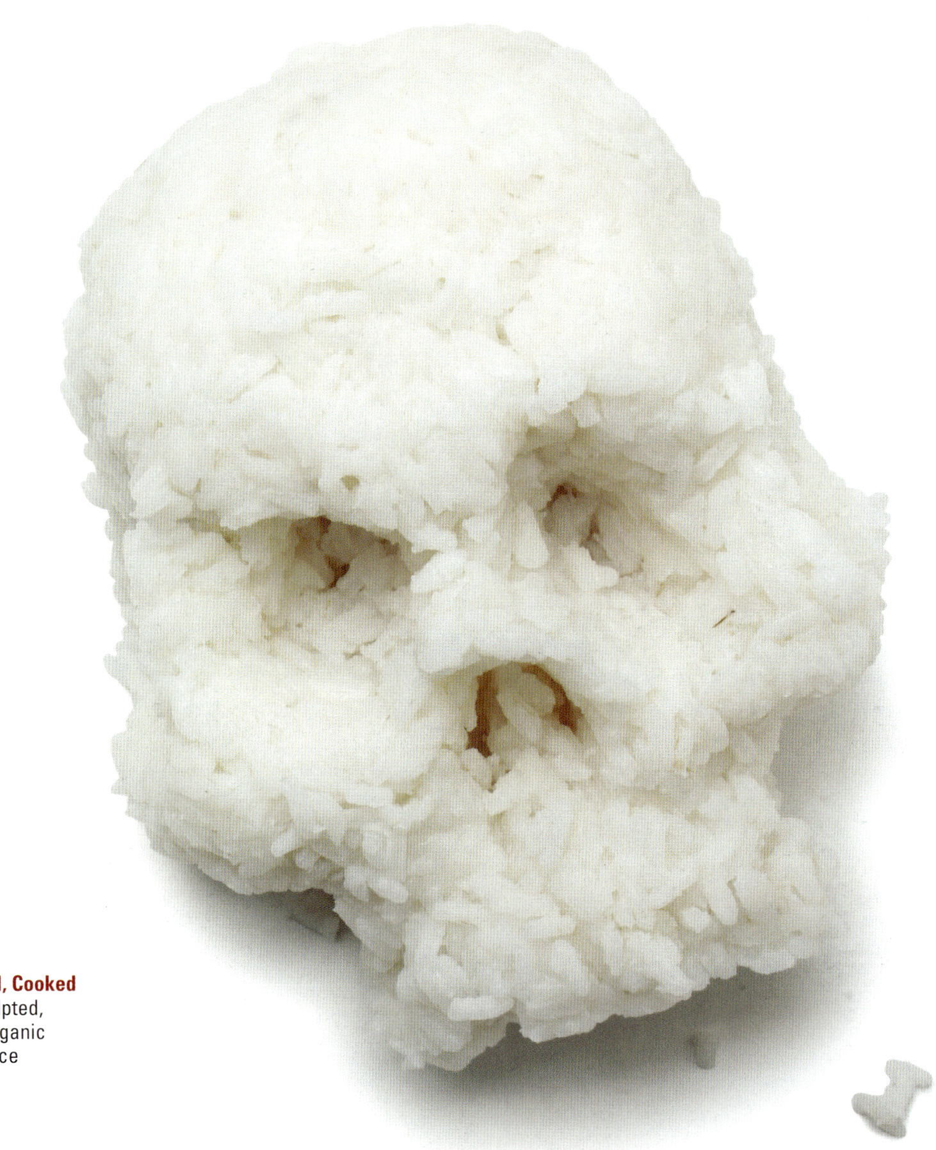

Rice Skull, Cooked
Hand-sculpted, cooked organic basmati rice

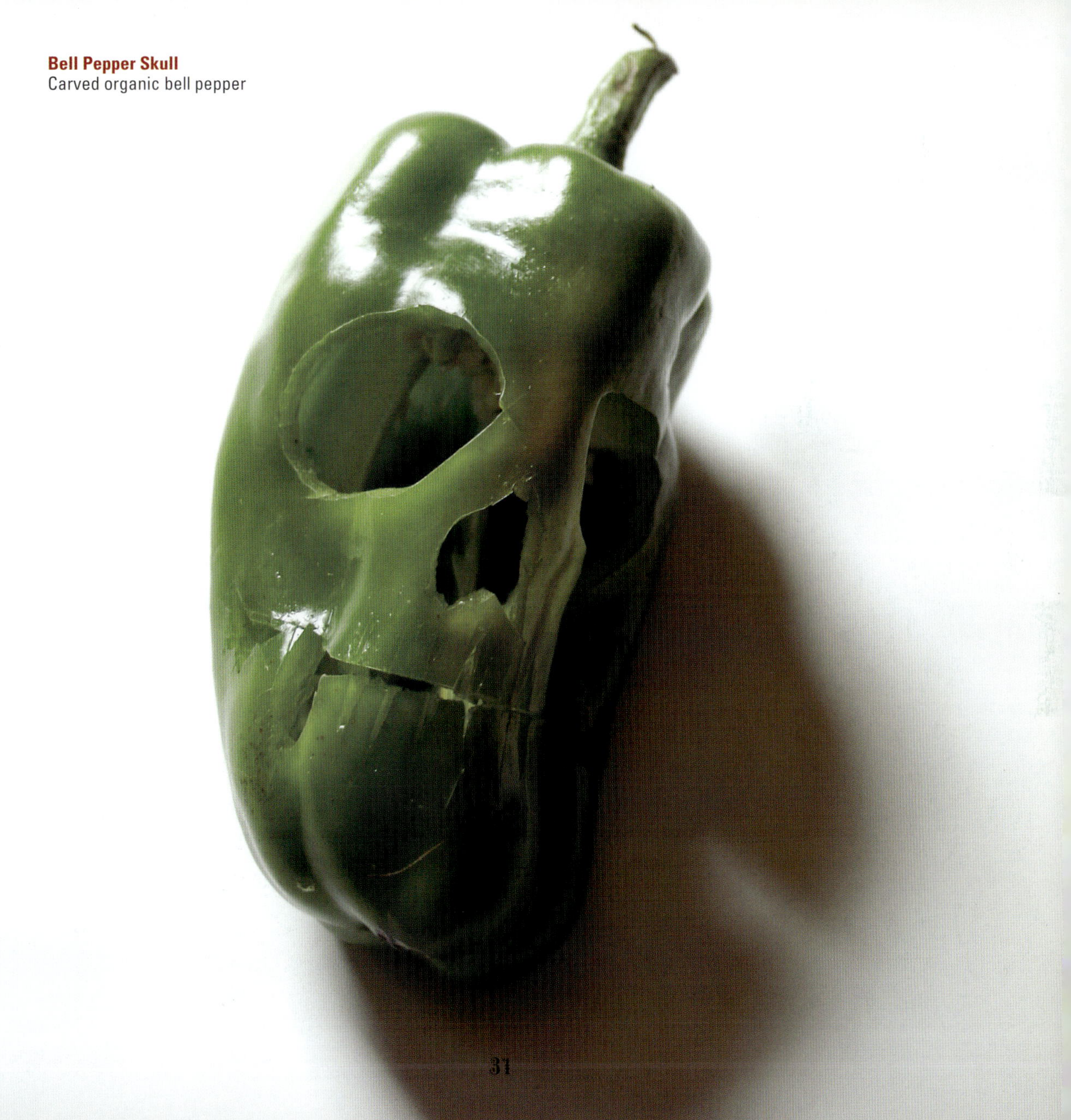

Bell Pepper Skull
Carved organic bell pepper

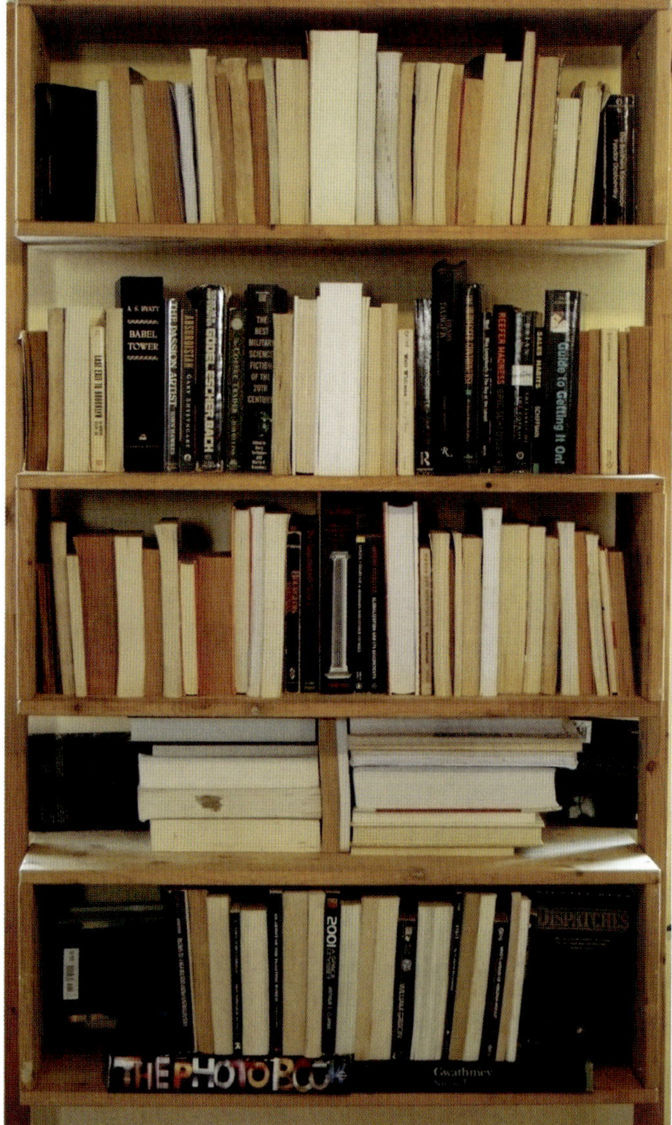

Bookshelf Skull
Arranged books. (This is a friend's bookshelf, I left it this way after visiting for a few days)

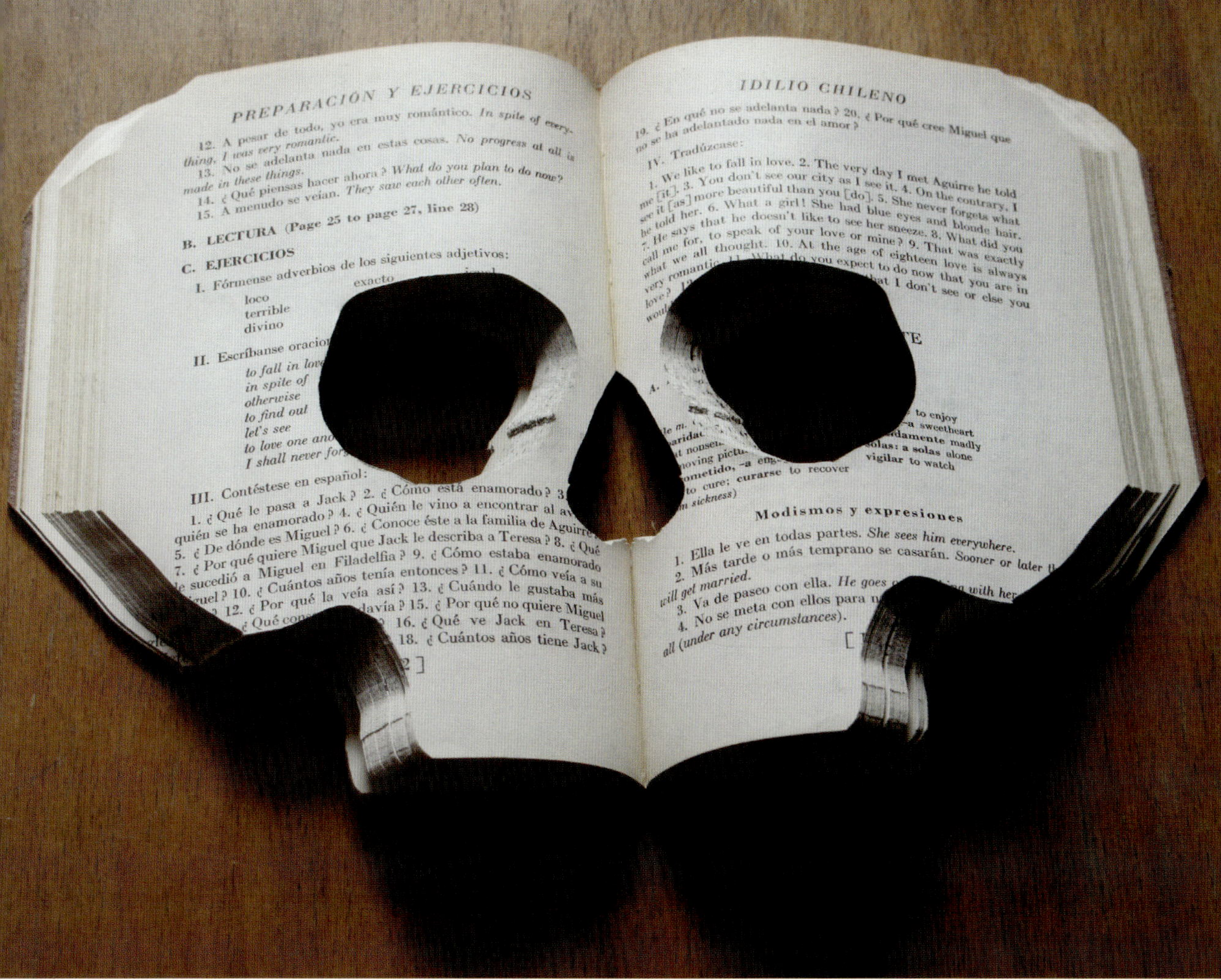

Libro del Cráneo Book cut with scroll saw. (I swear I have nothing personal against books.)

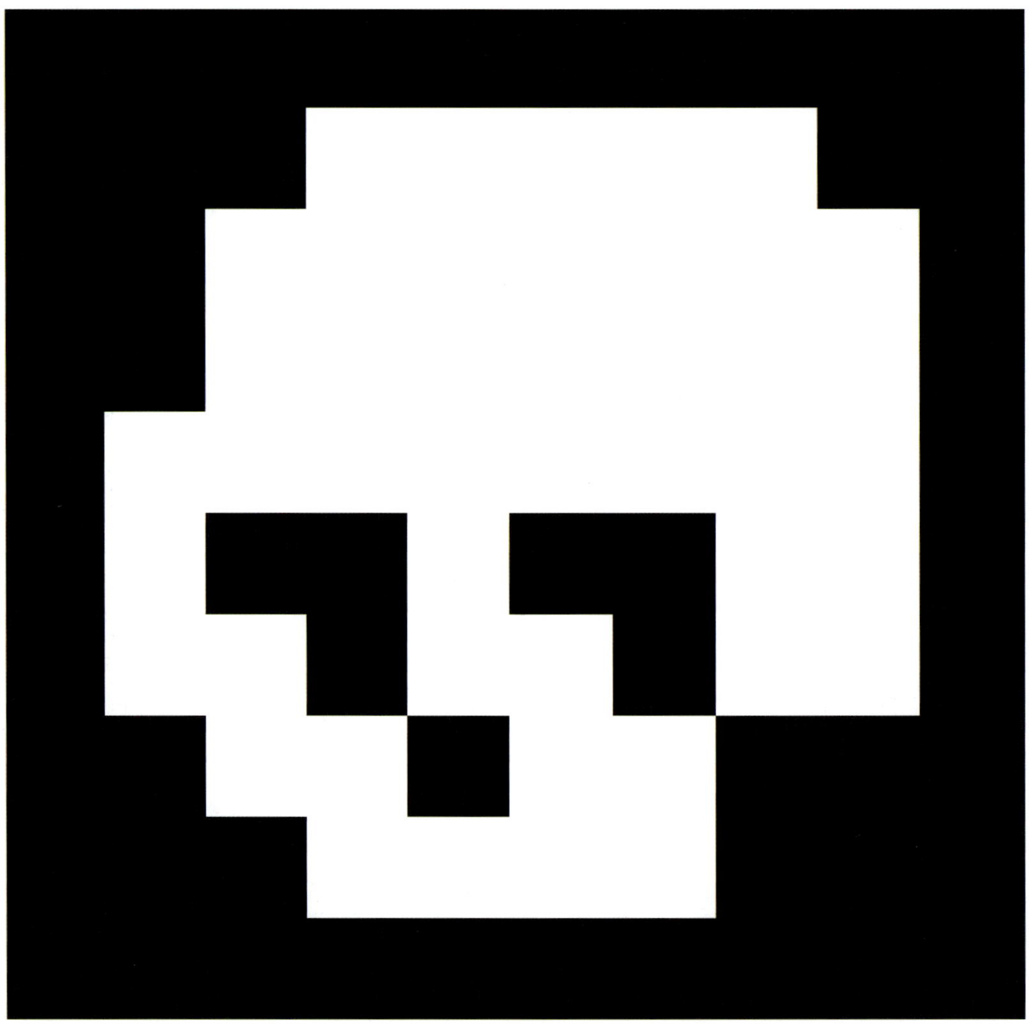

8 x 8 Skull Image made in 8 pixel by 8 pixel grid and reproduced in recycled sticky notes

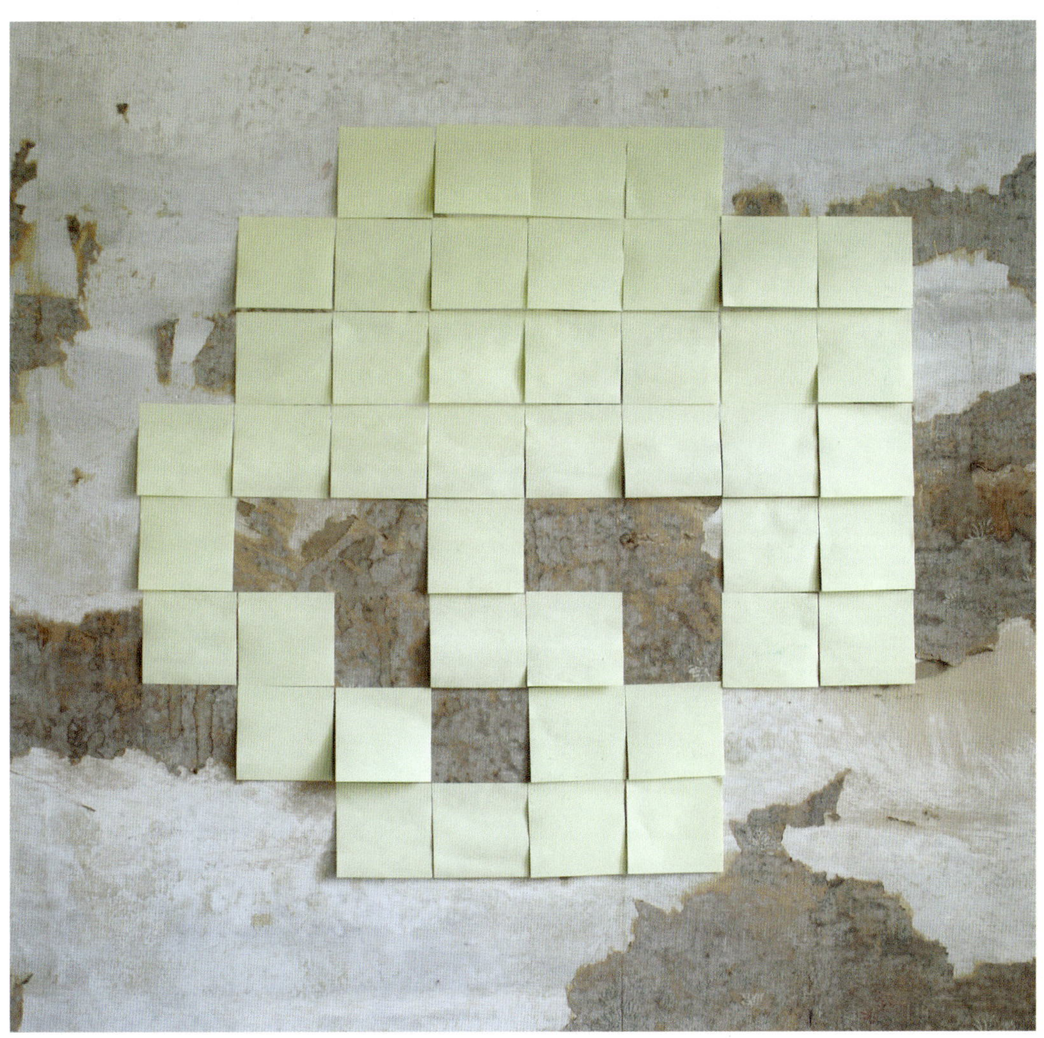

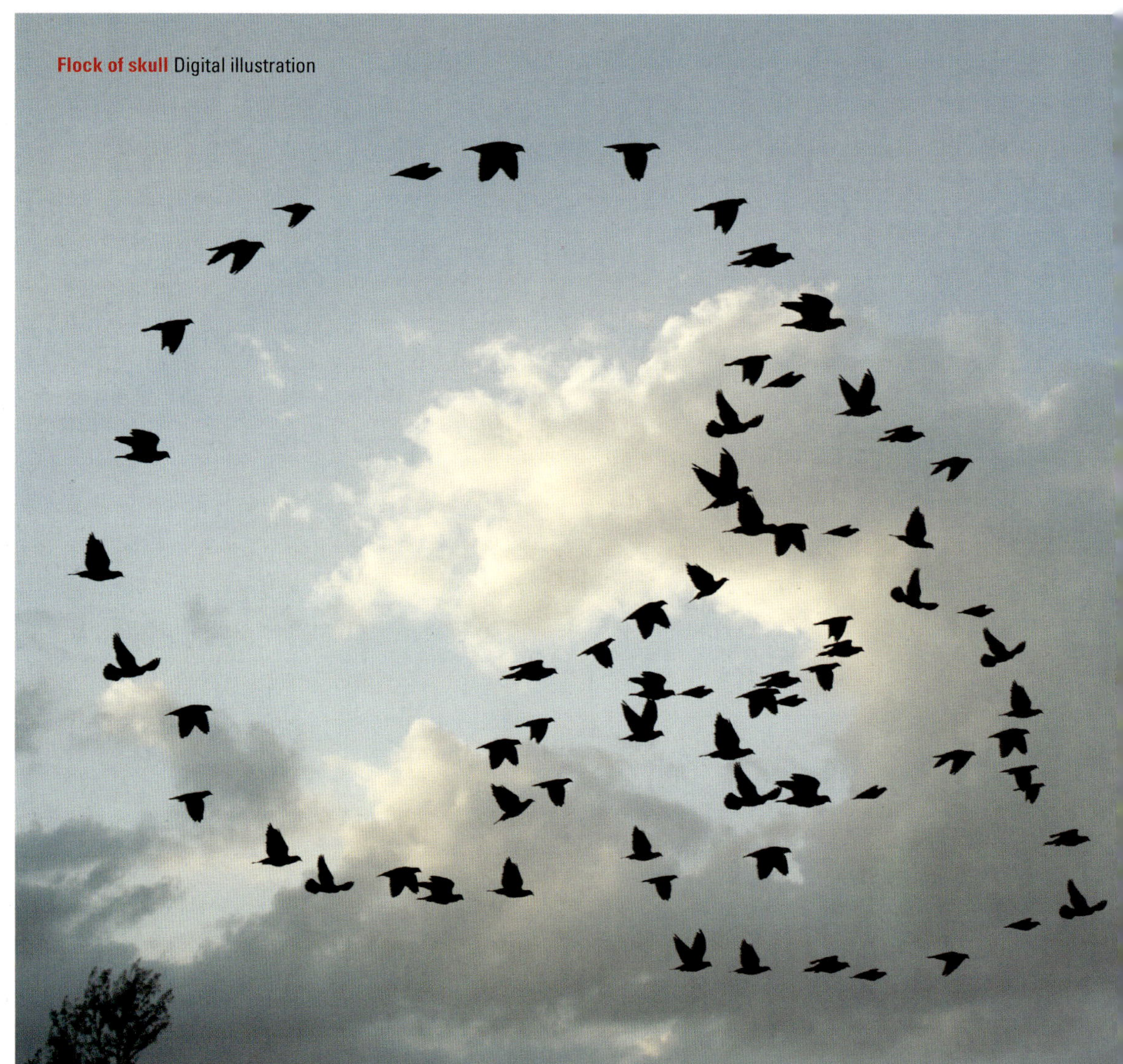

Flock of skull Digital illustration

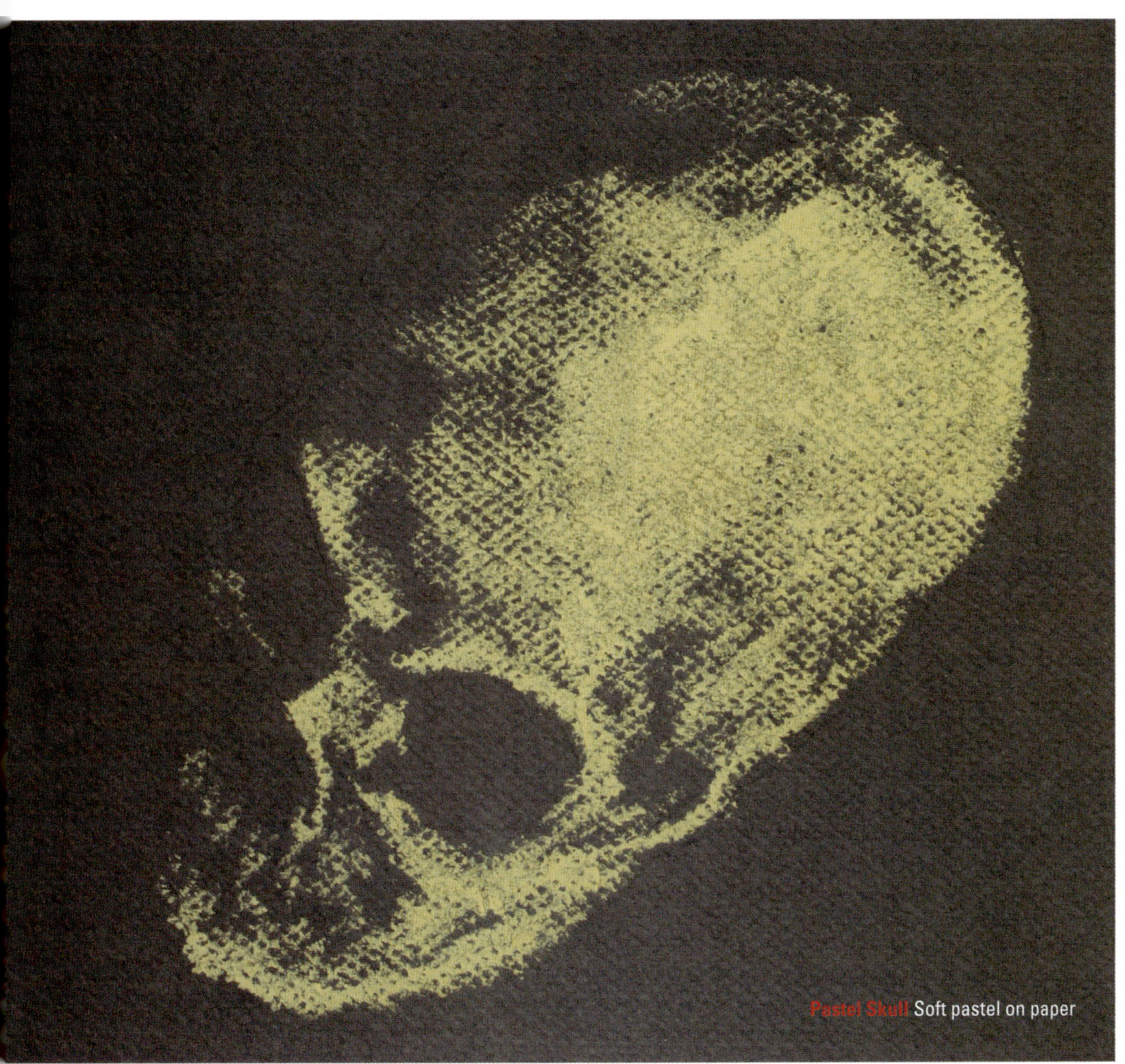

Pastel Skull Soft pastel on paper

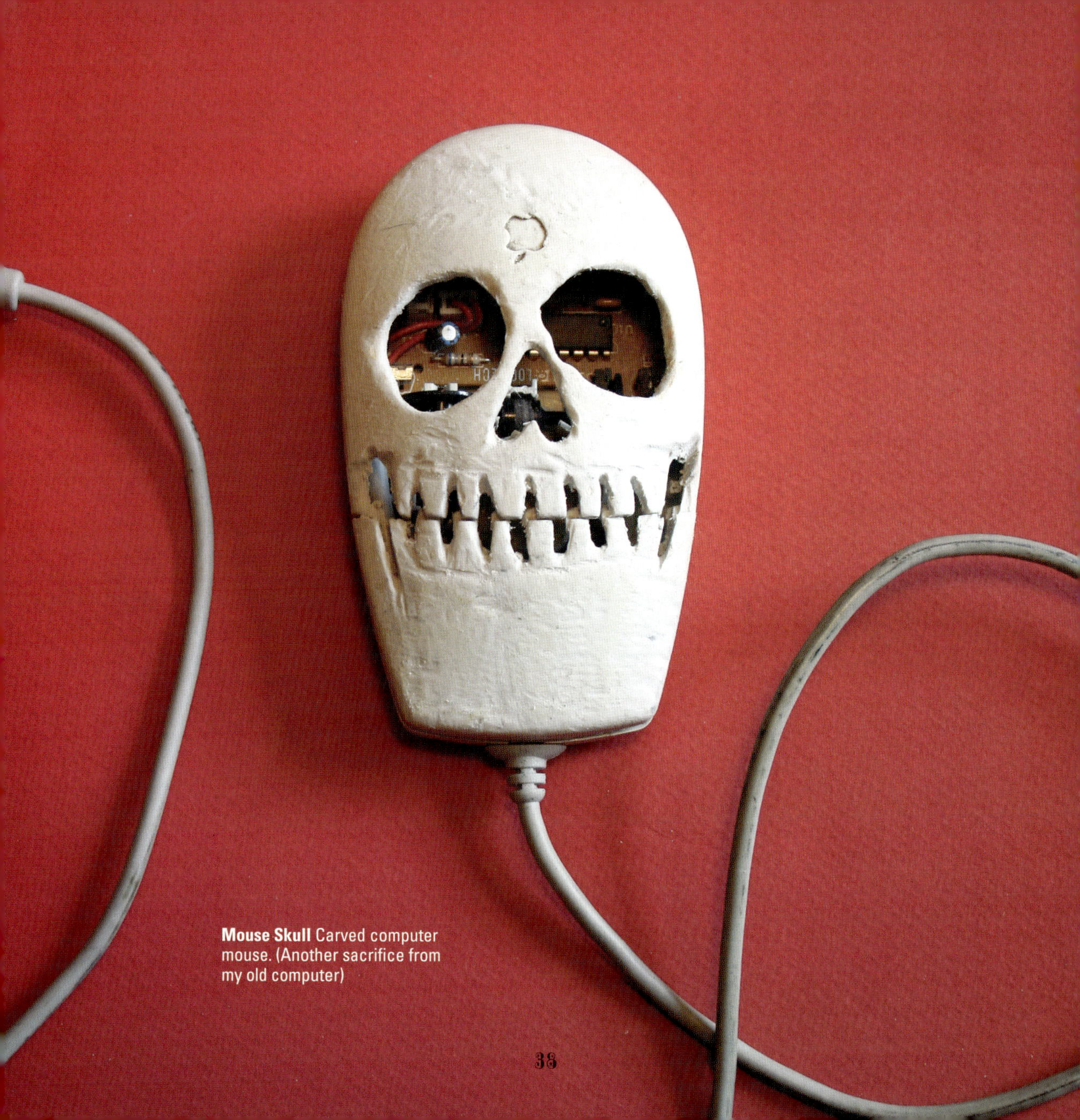

Mouse Skull Carved computer mouse. (Another sacrifice from my old computer)

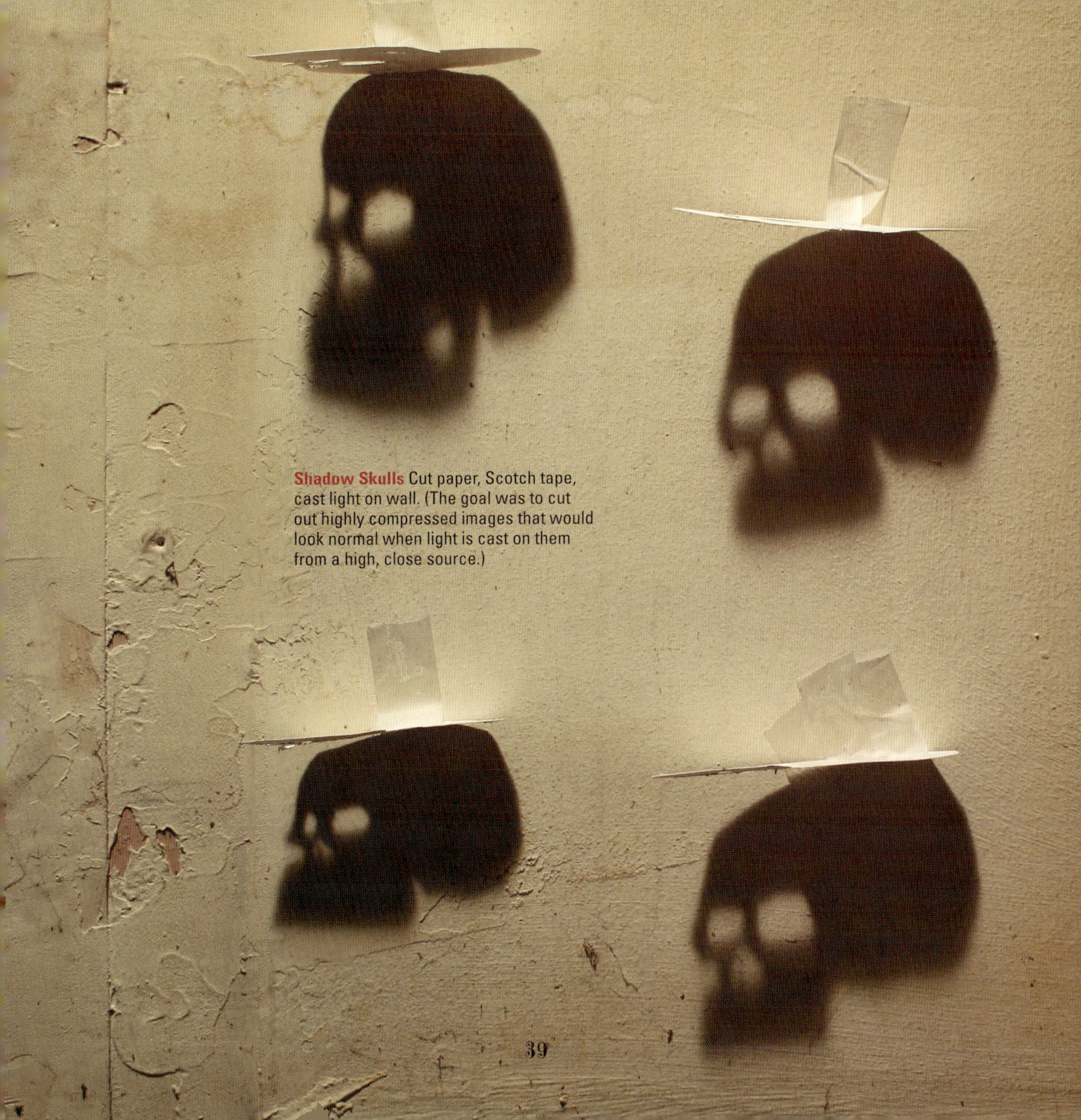

Shadow Skulls Cut paper, Scotch tape, cast light on wall. (The goal was to cut out highly compressed images that would look normal when light is cast on them from a high, close source.)

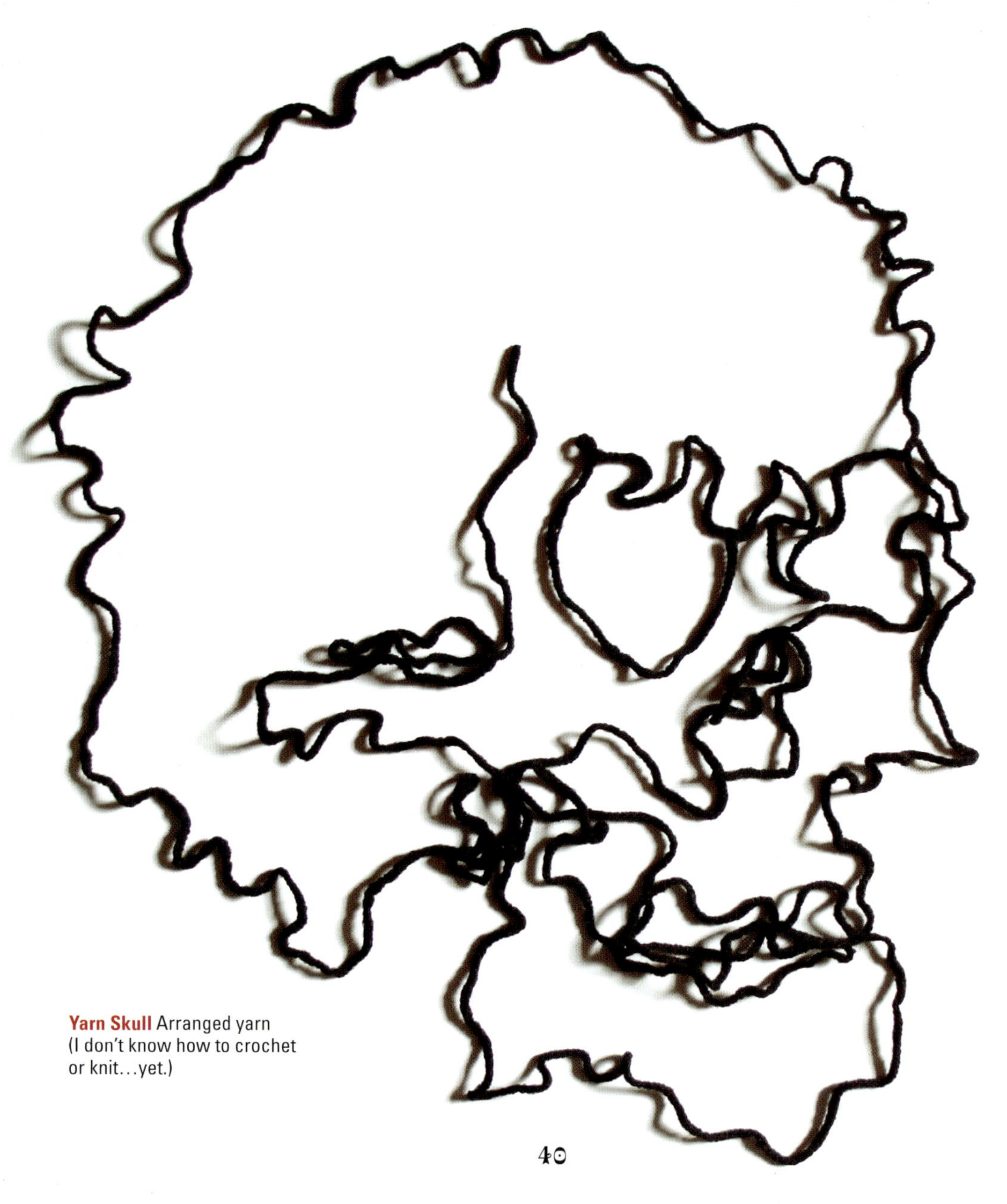

Yarn Skull Arranged yarn
(I don't know how to crochet
or knit…yet.)

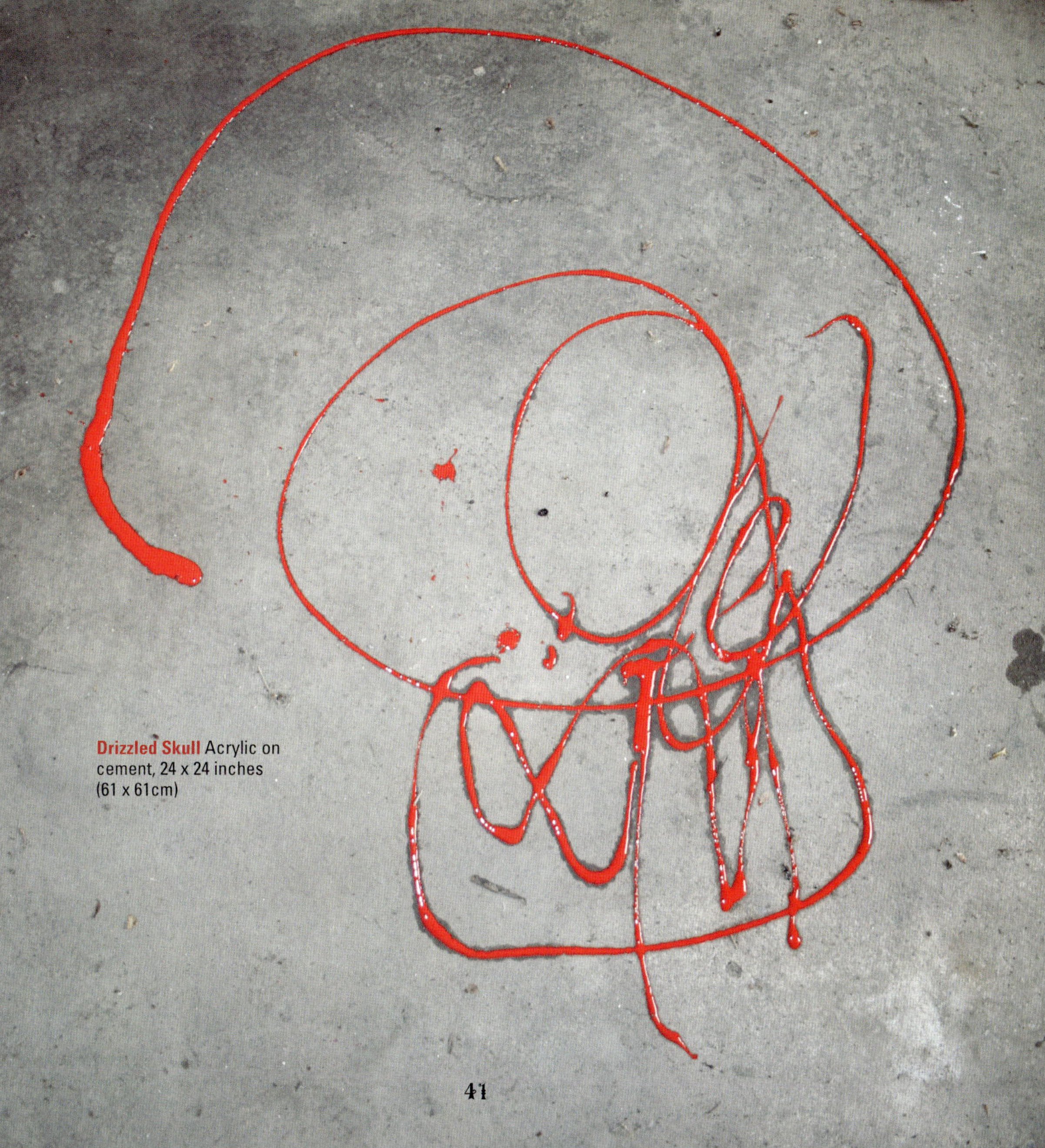

Drizzled Skull Acrylic on cement, 24 x 24 inches (61 x 61cm)

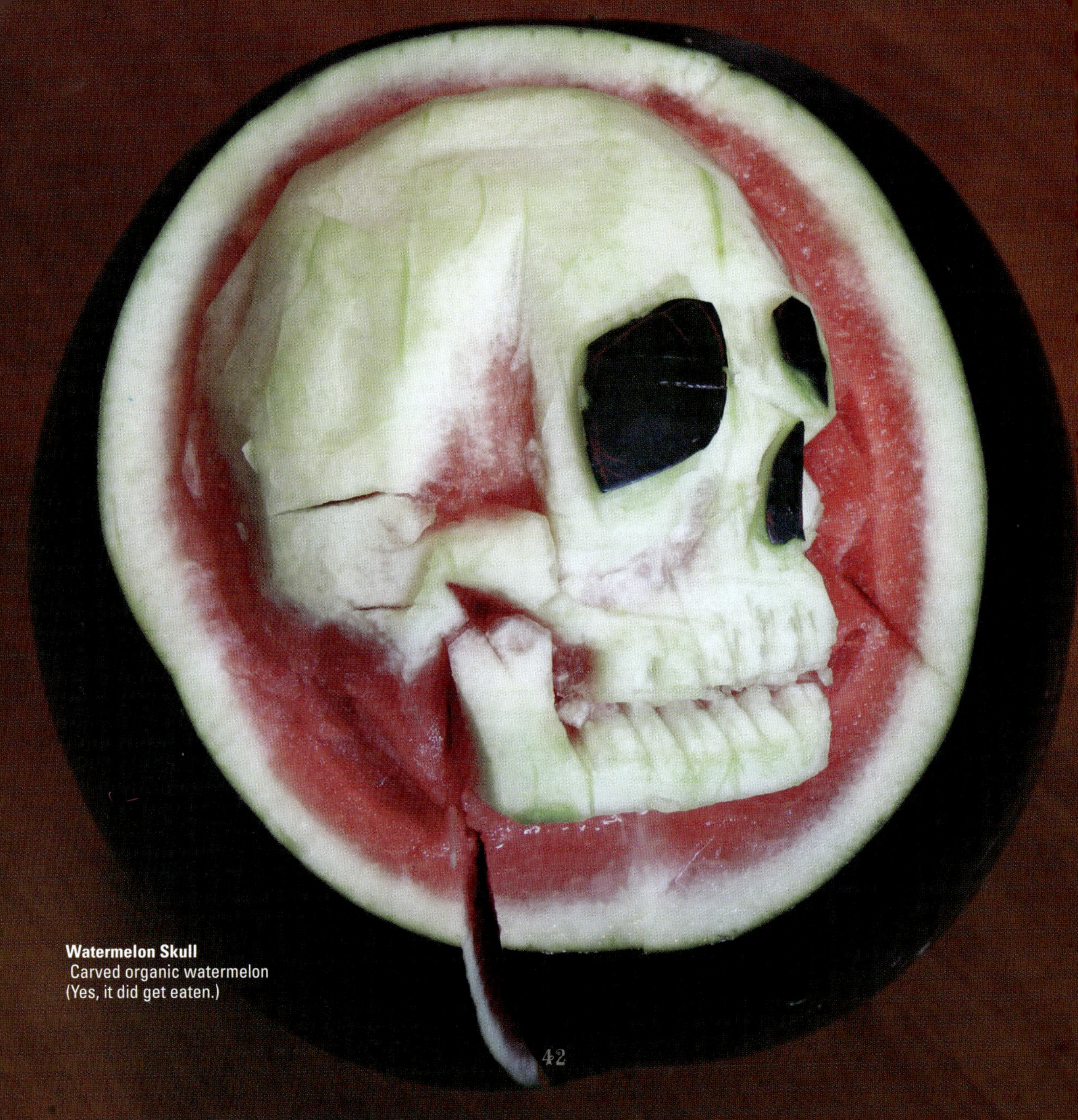

Watermelon Skull
Carved organic watermelon
(Yes, it did get eaten.)

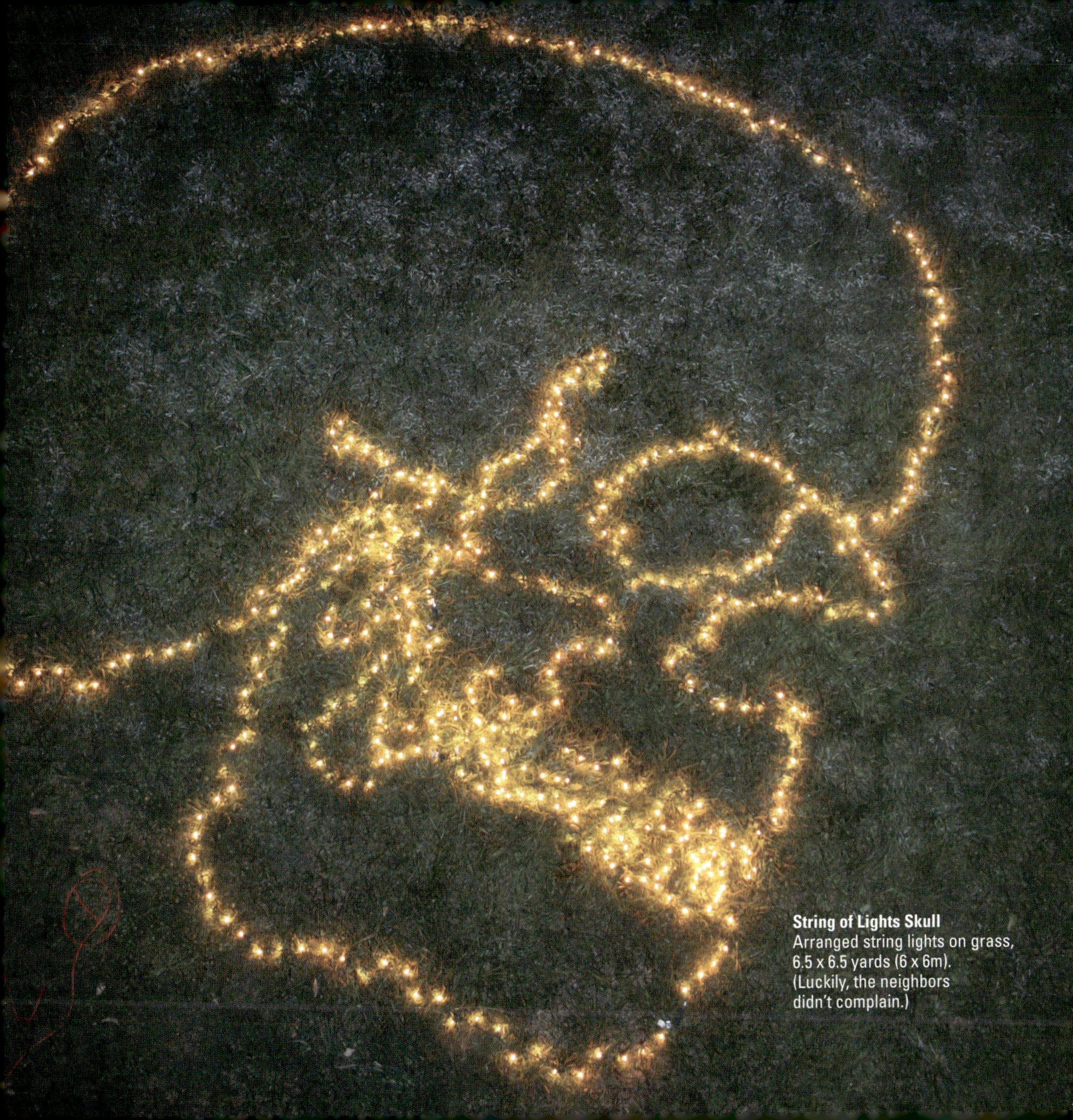

String of Lights Skull
Arranged string lights on grass, 6.5 x 6.5 yards (6 x 6m). (Luckily, the neighbors didn't complain.)

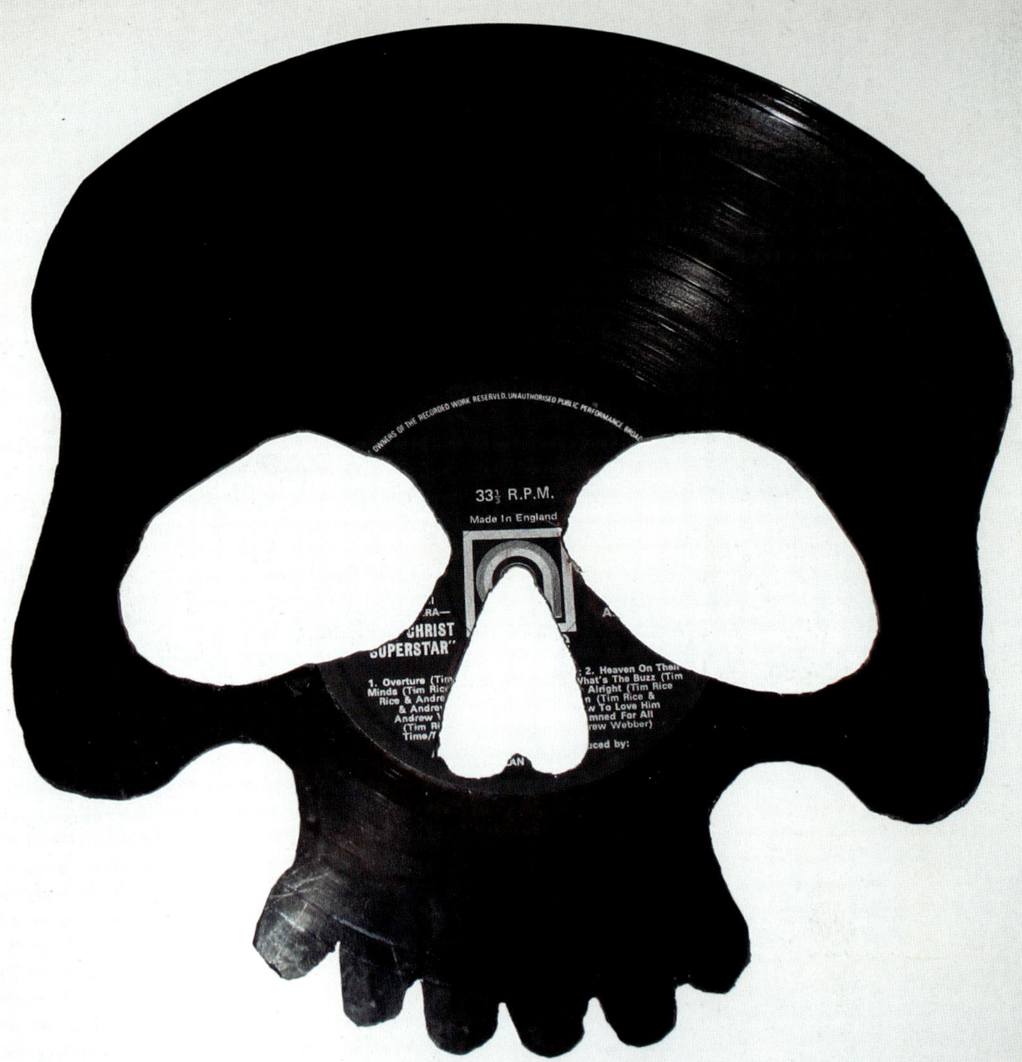

Vinyl Skull Cut vinyl LP

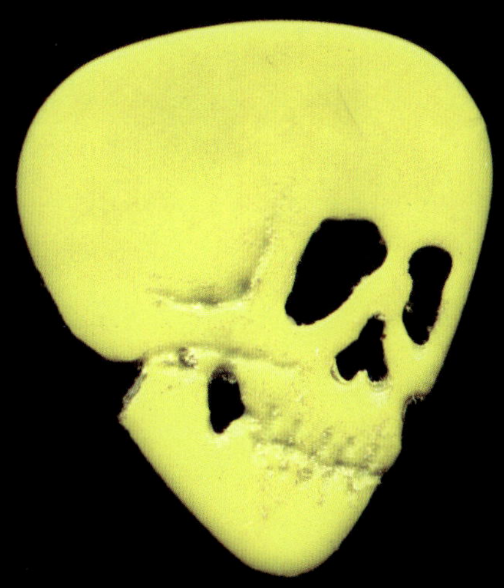

Skull Pick Carved, found guitar pick

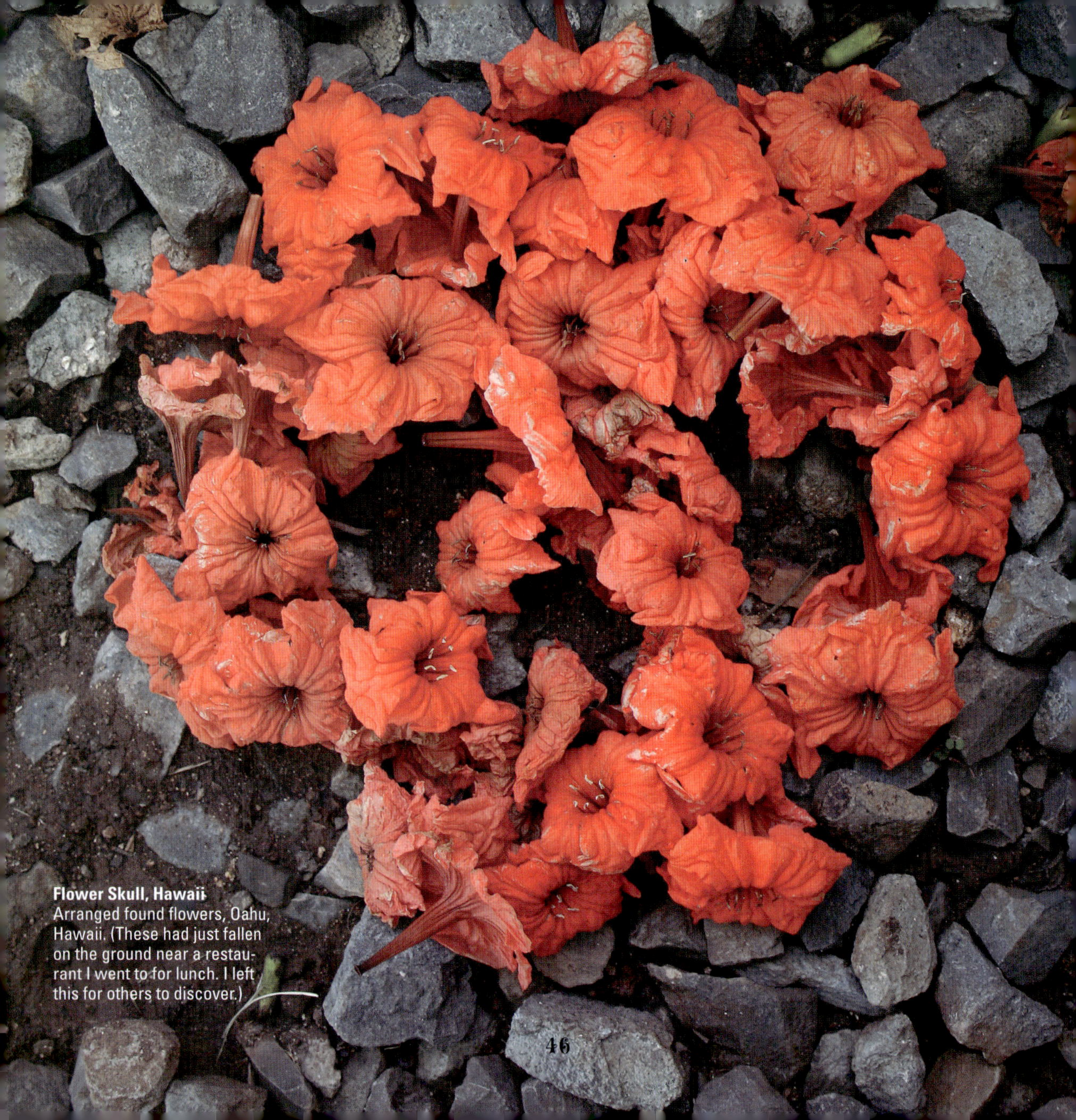

Flower Skull, Hawaii
Arranged found flowers, Oahu, Hawaii. (These had just fallen on the ground near a restaurant I went to for lunch. I left this for others to discover.)

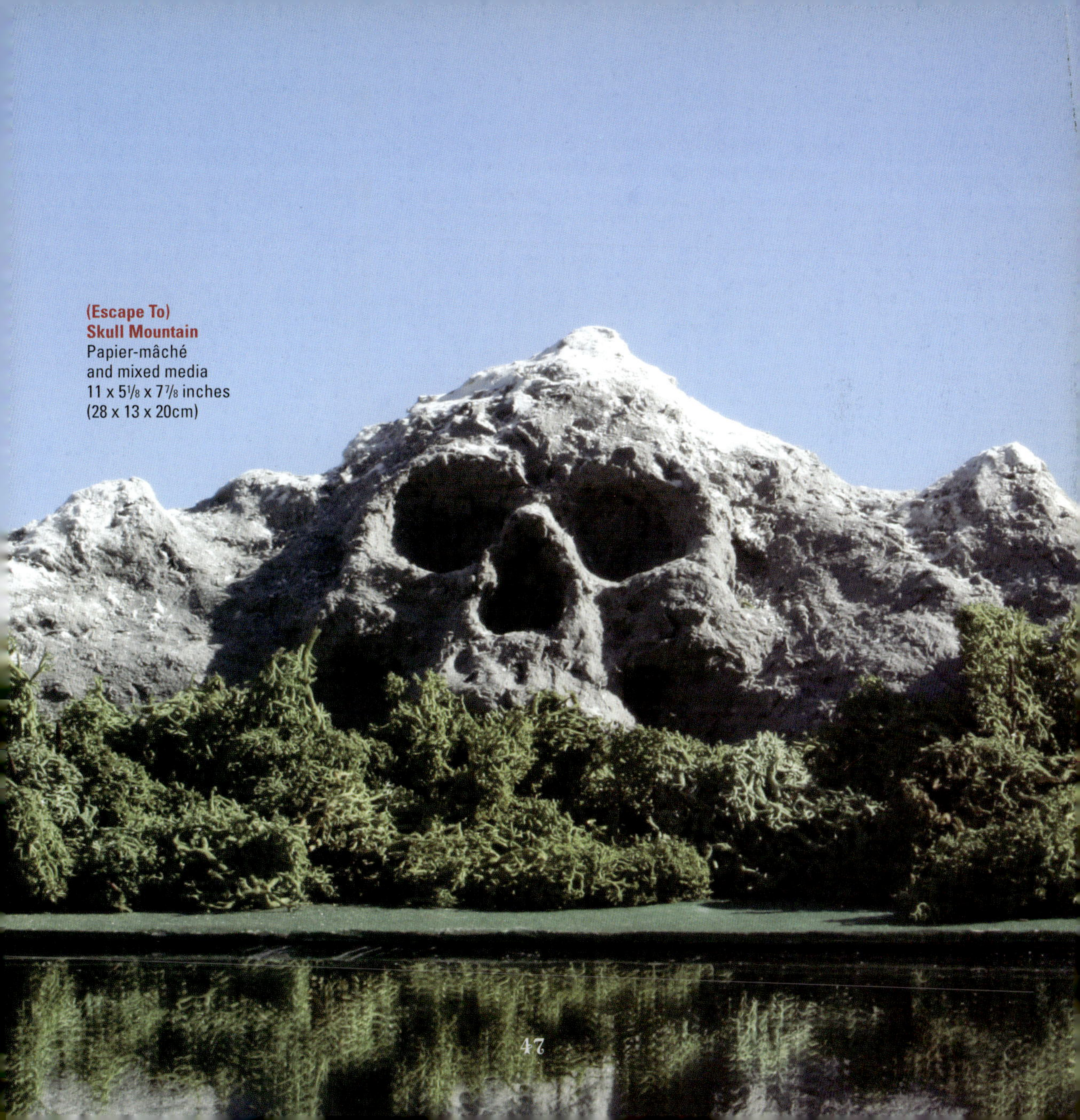

(Escape To) Skull Mountain
Papier-mâché
and mixed media
11 x $5^{1}/_{8}$ x $7^{7}/_{8}$ inches
(28 x 13 x 20cm)

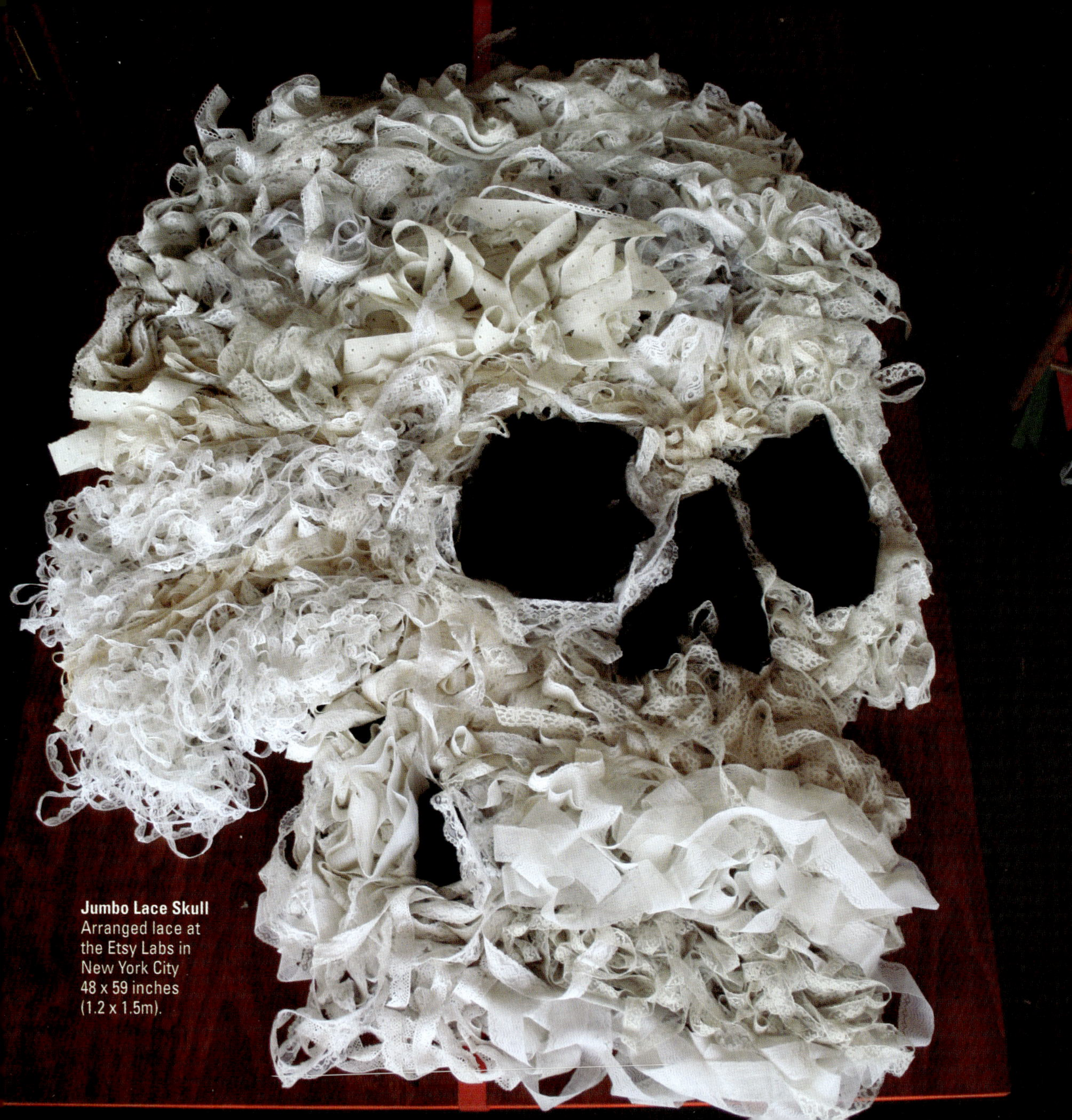

Jumbo Lace Skull
Arranged lace at the Etsy Labs in New York City
48 x 59 inches (1.2 x 1.5m).

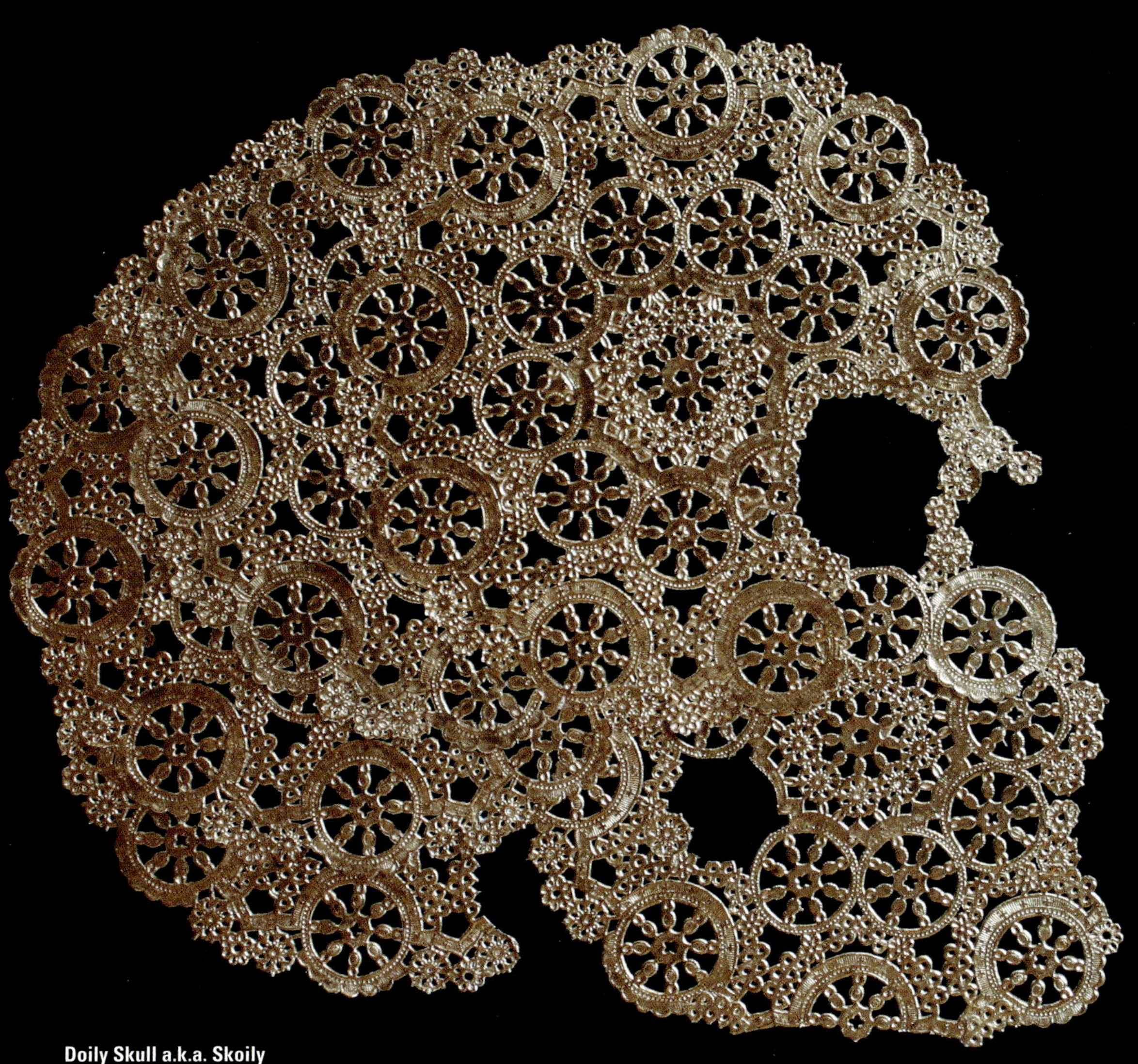

Doily Skull a.k.a. Skoily
Cut and arranged paper doilies,
13 x 11 inches (33 x 28cm)

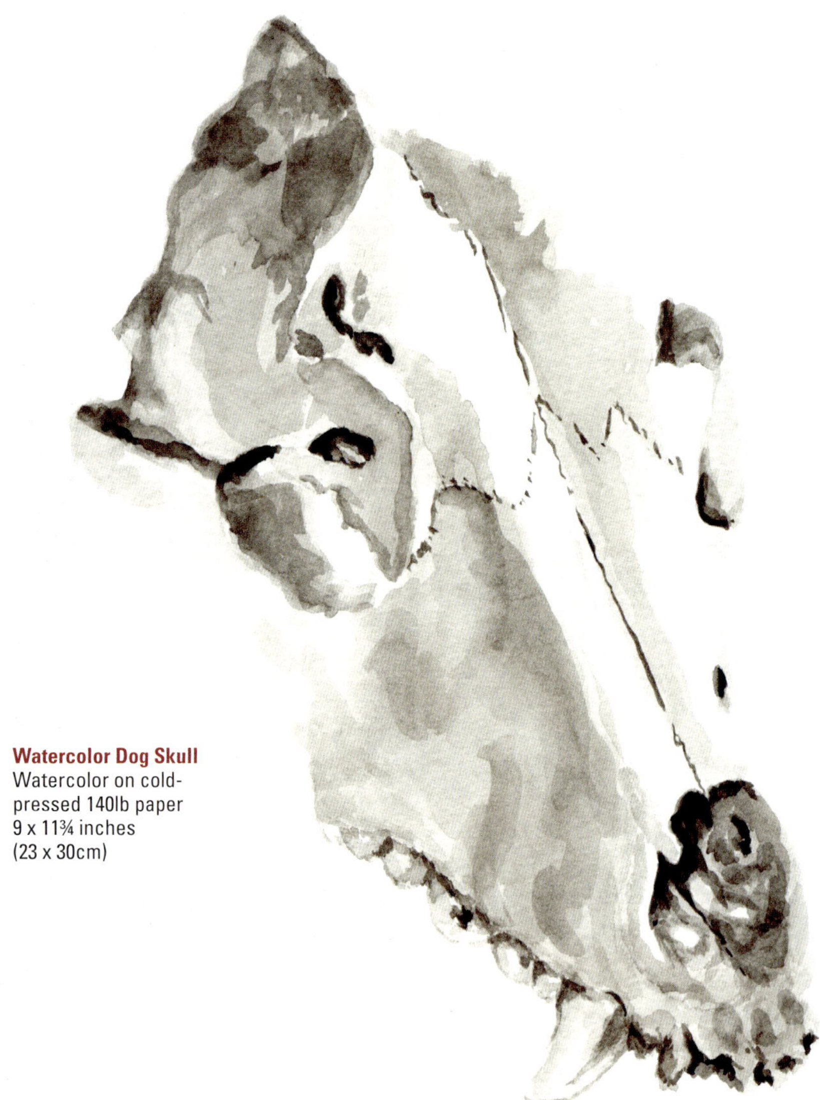

Watercolor Dog Skull
Watercolor on cold-pressed 140lb paper
9 x 11¾ inches
(23 x 30cm)

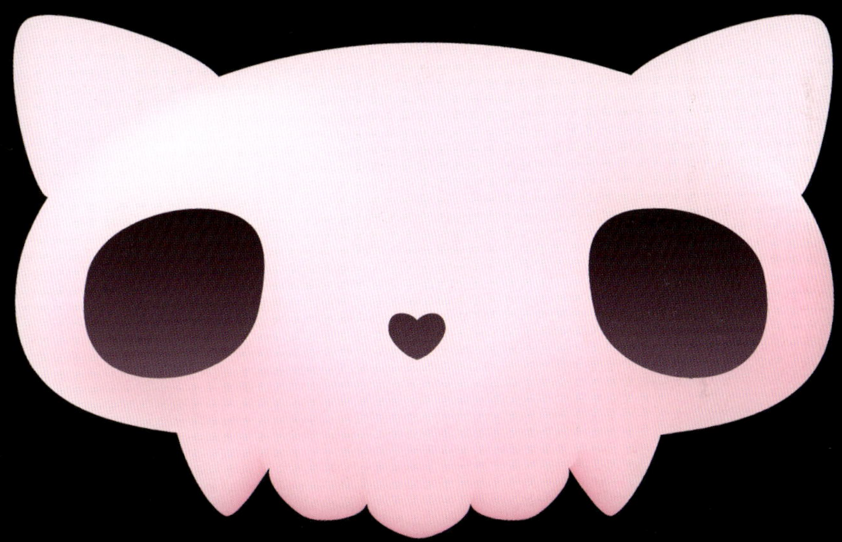

Cutest Skull…Ever! Digital illustration. My limited online research indicated the following requirements for ultimate cuteness: 1. Baby Animals (preferably kittens)
2. Japanese Stuff (especially cartoons)
3. The Color Pink
4. Rounded Corners

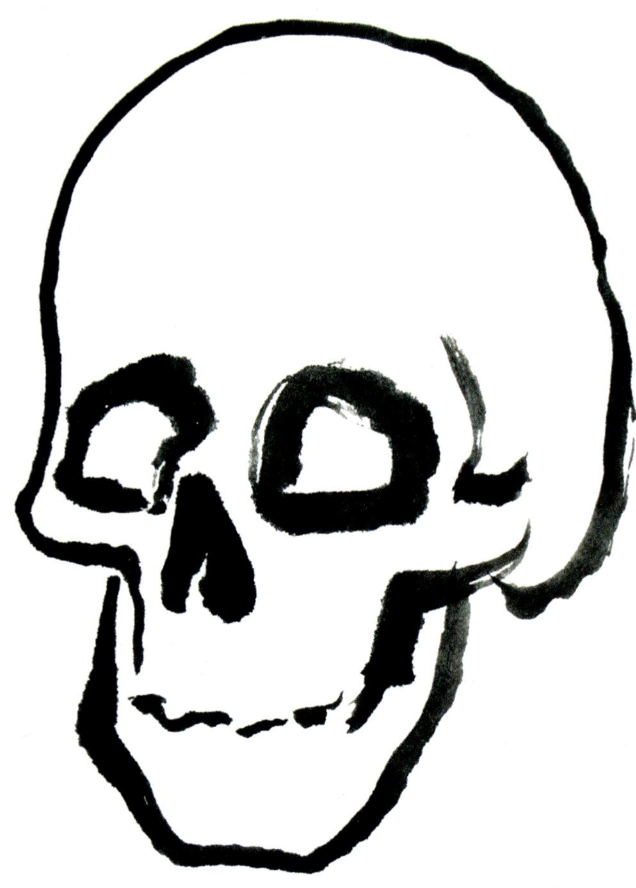

Sumi-e Skull Hand-ground ink on rice paper, with chop signature. (Having never worked with this medium before, it took 45 attempts to get just a few that I liked.)

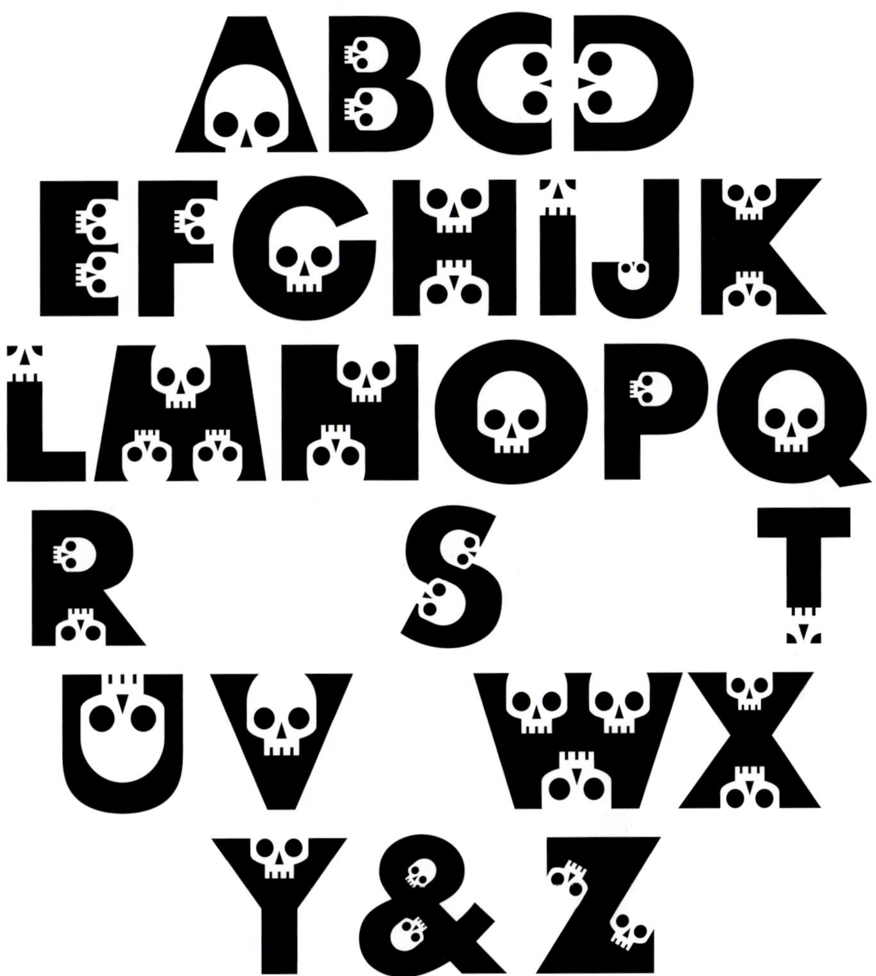

Skullphabet #1 Display capital letterforms based on Futura Bold. (Download this font for free at www.skulladay.com!)

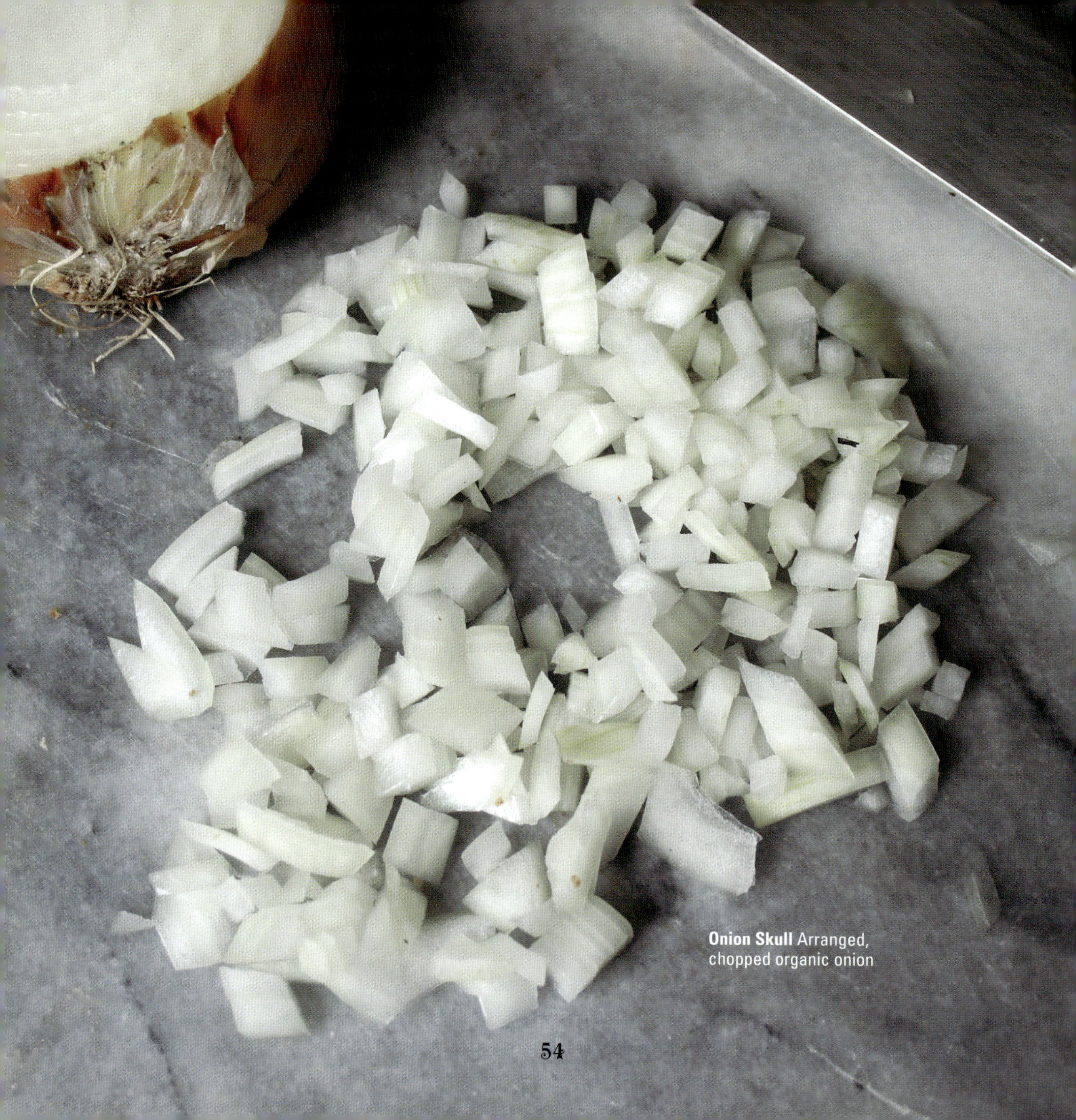

Onion Skull Arranged, chopped organic onion

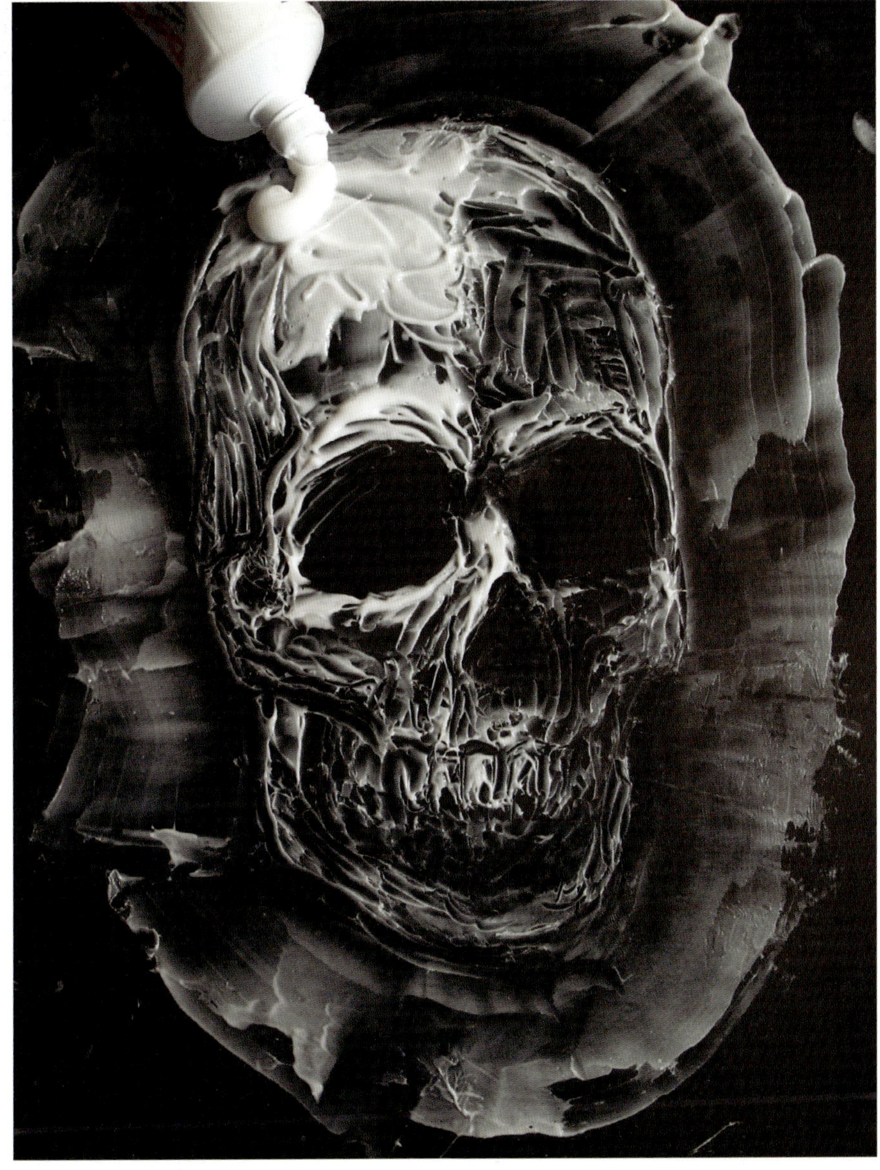

(Minty Fresh) Skullpaste Knife-spread toothpaste

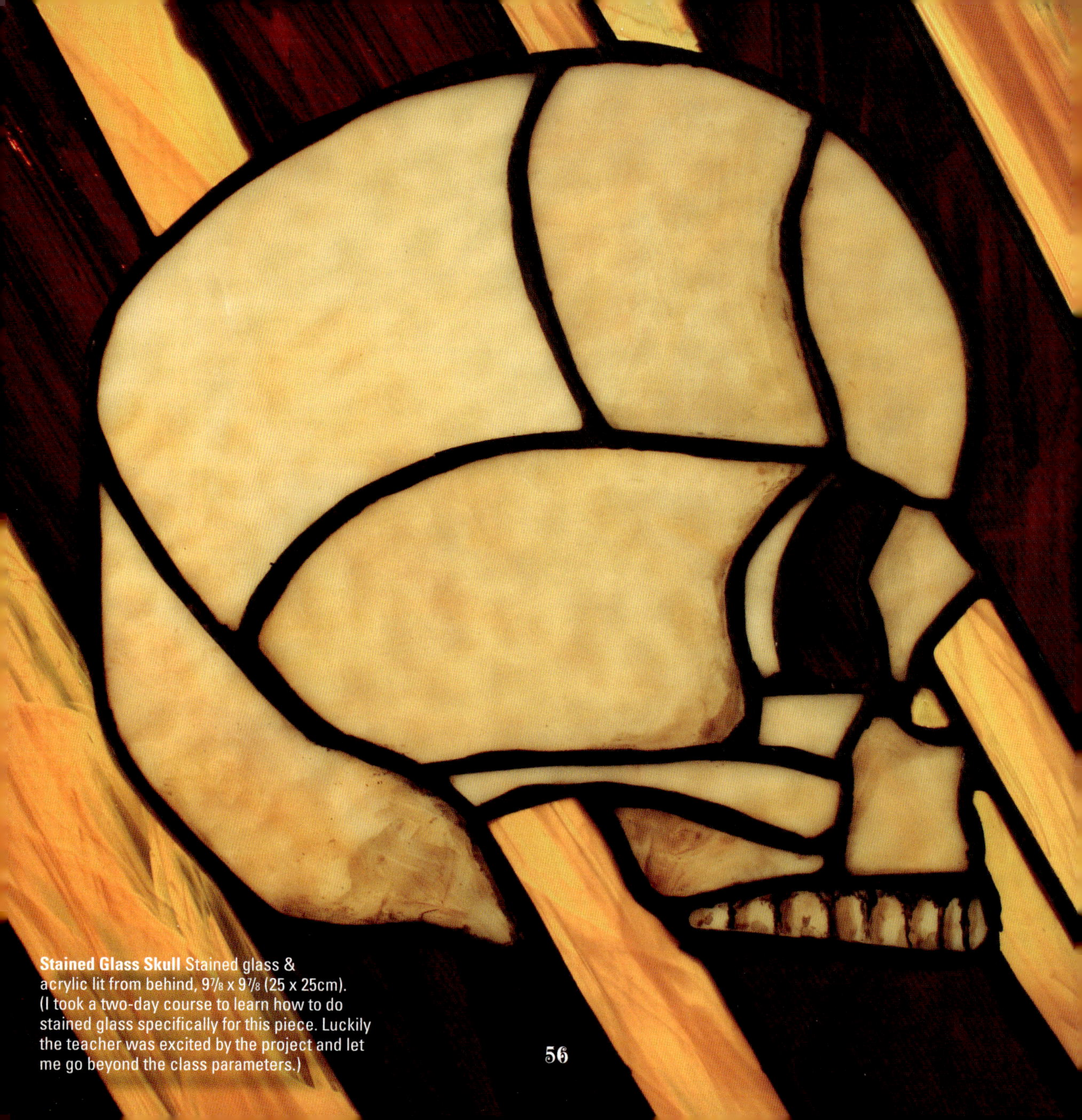

Stained Glass Skull Stained glass & acrylic lit from behind, 9⅞ x 9⅞ (25 x 25cm). (I took a two-day course to learn how to do stained glass specifically for this piece. Luckily the teacher was excited by the project and let me go beyond the class parameters.)

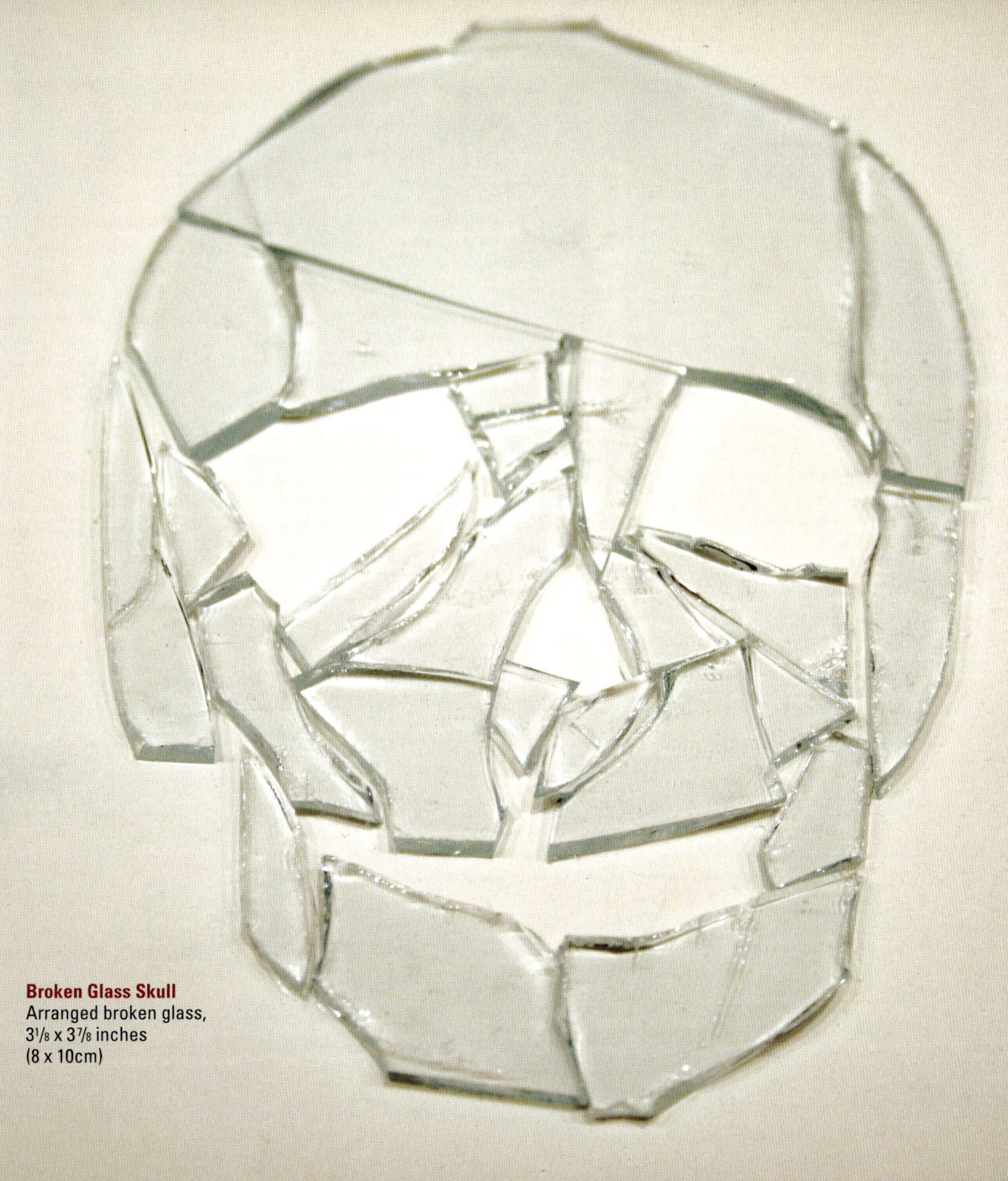

Broken Glass Skull
Arranged broken glass,
3 1/8 x 3 7/8 inches
(8 x 10cm)

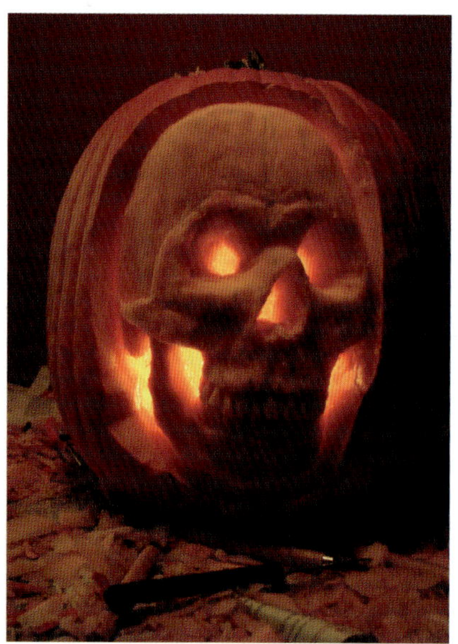

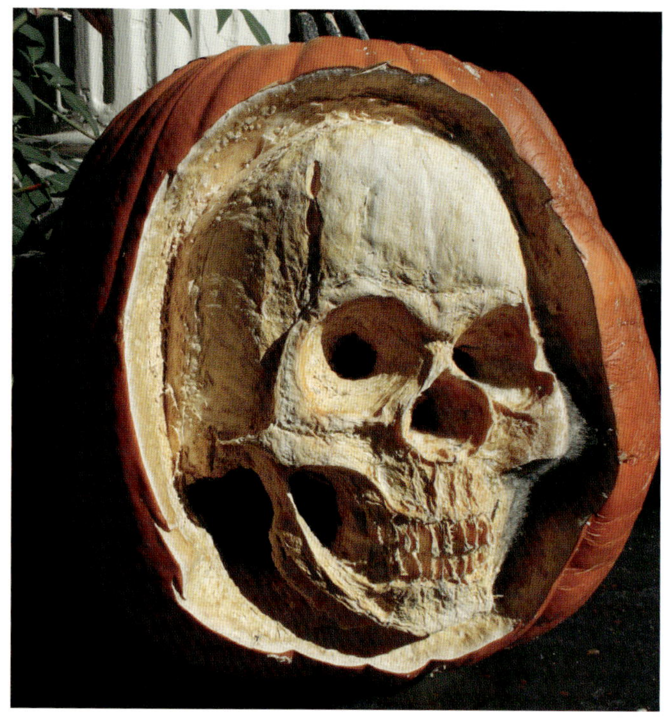

Skull-'O-Lantern Carved pumpkin, lit, unlit, and (middle photo) 10 days later. (The pumpkin skull finally collapsed into a pile of goop on my stoop and was ceremoniously dumped in my compost pile.)

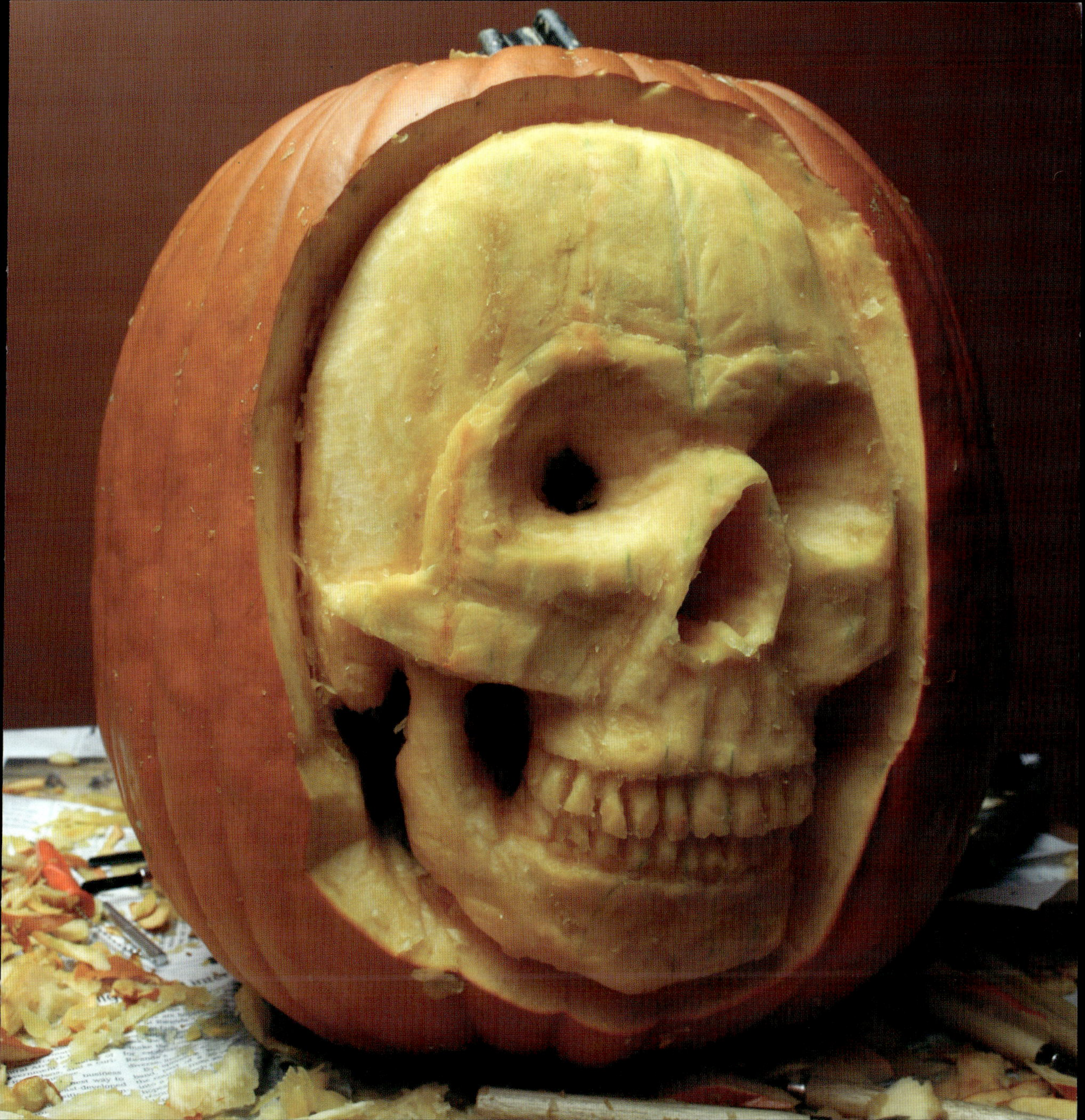

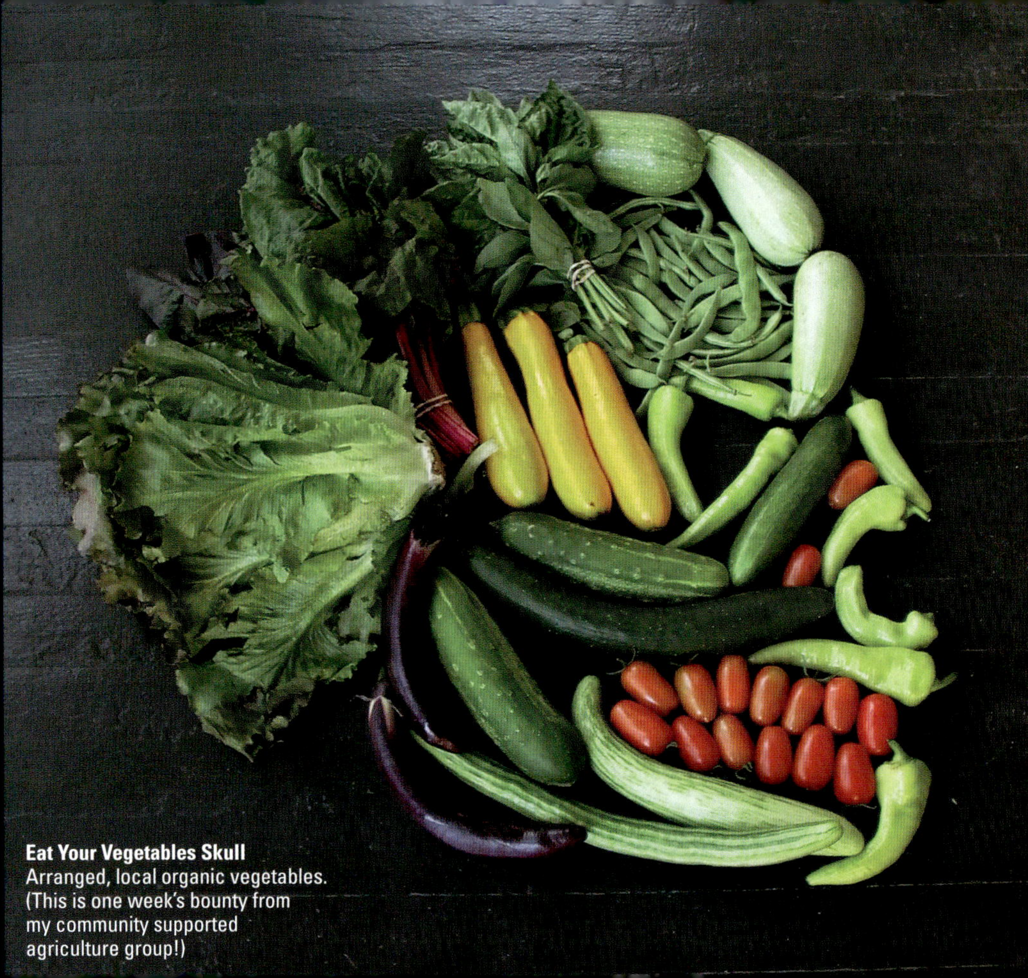

Eat Your Vegetables Skull
Arranged, local organic vegetables.
(This is one week's bounty from
my community supported
agriculture group!)

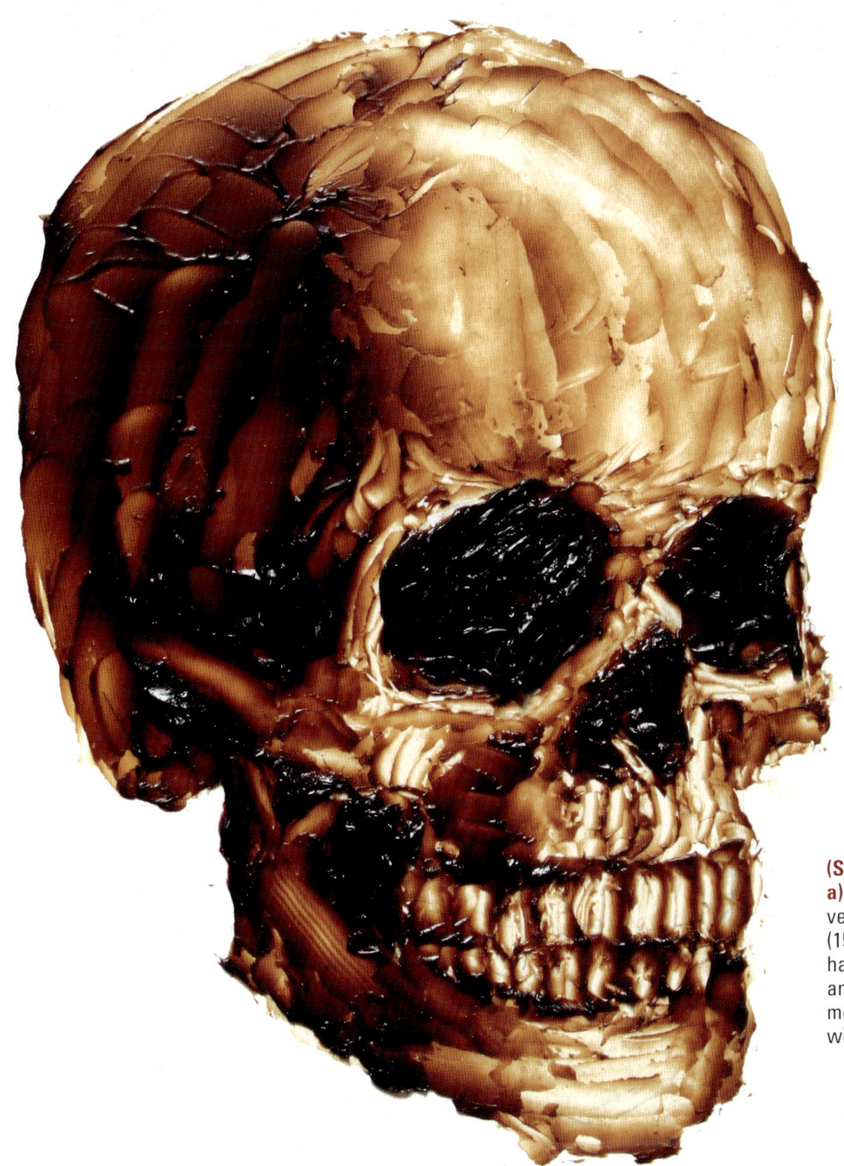

(She Just Smiled and Gave Me a) Vegemite Skull Knife-spread vegemite 5⅞ x 7½ inches (15 x 19cm). (I mentioned never having tried Vegemite on my site, and an Australian fan kindly sent me a jar. Not only did I make this with it, but I actually ate some!)

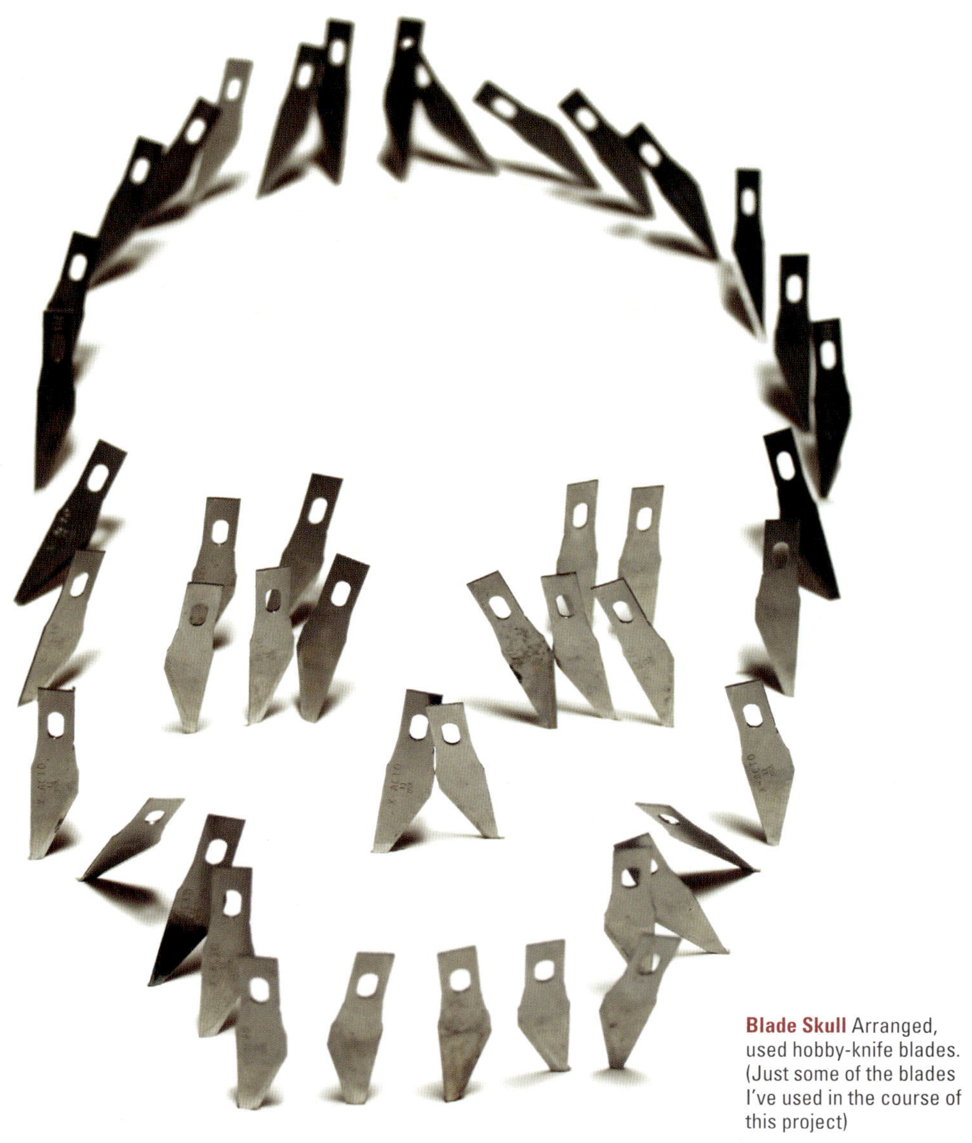

Blade Skull Arranged, used hobby-knife blades. (Just some of the blades I've used in the course of this project)

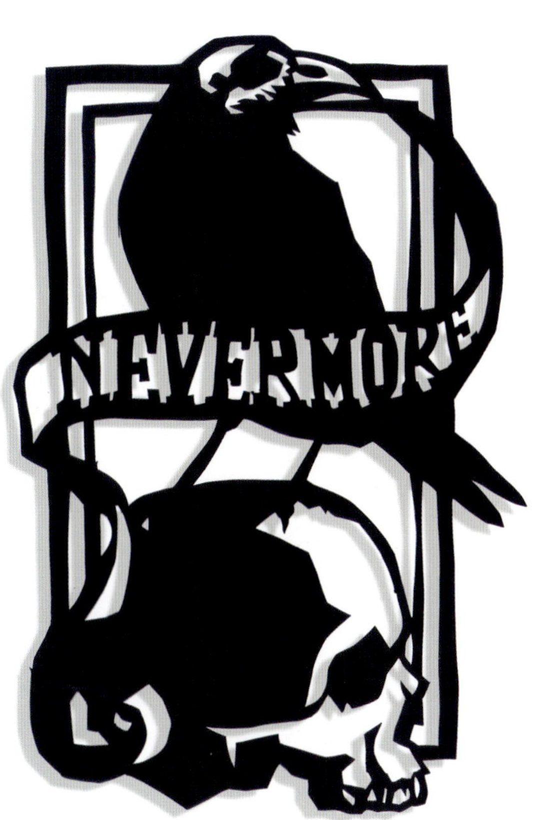

Quoth The Skull…
Hand-cut paper,
5 7/8 x 9 7/8 inches (15 x 25cm)

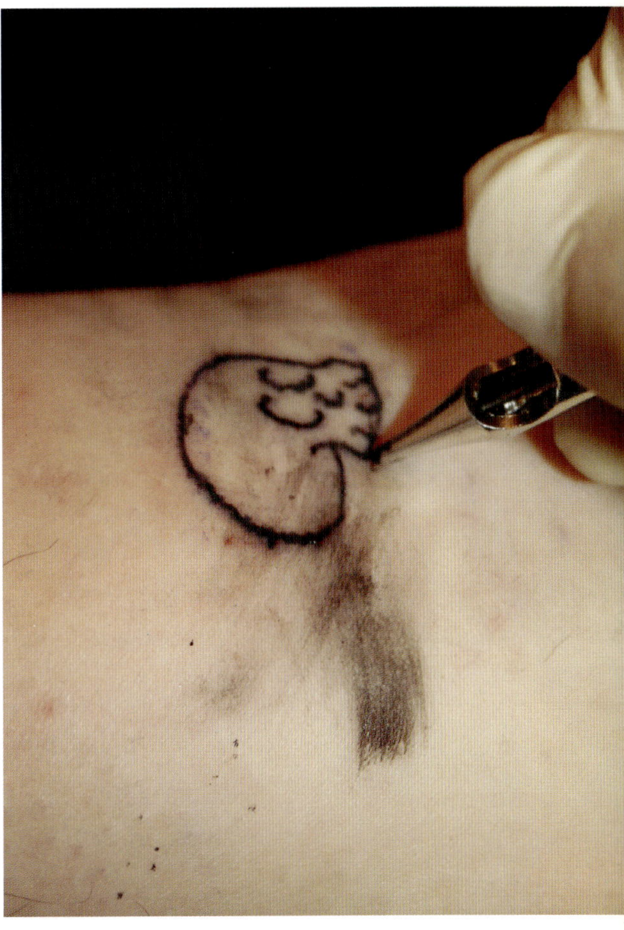

Tattoo Skull, Self-Inflicted:
Ink in skin. (Yes, this is a real tattoo that I really gave myself, using professional equipment.)

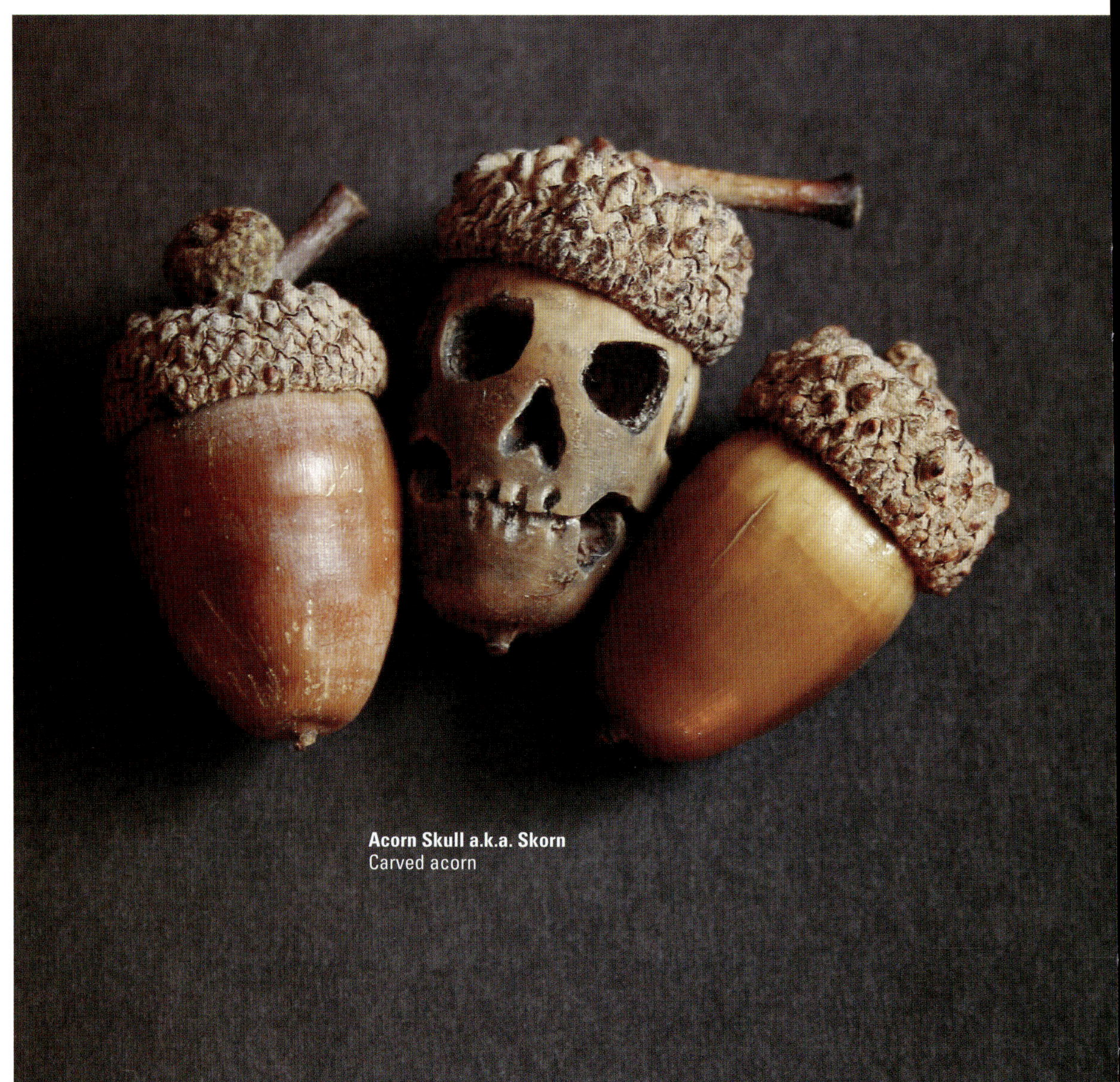

Acorn Skull a.k.a. Skorn
Carved acorn

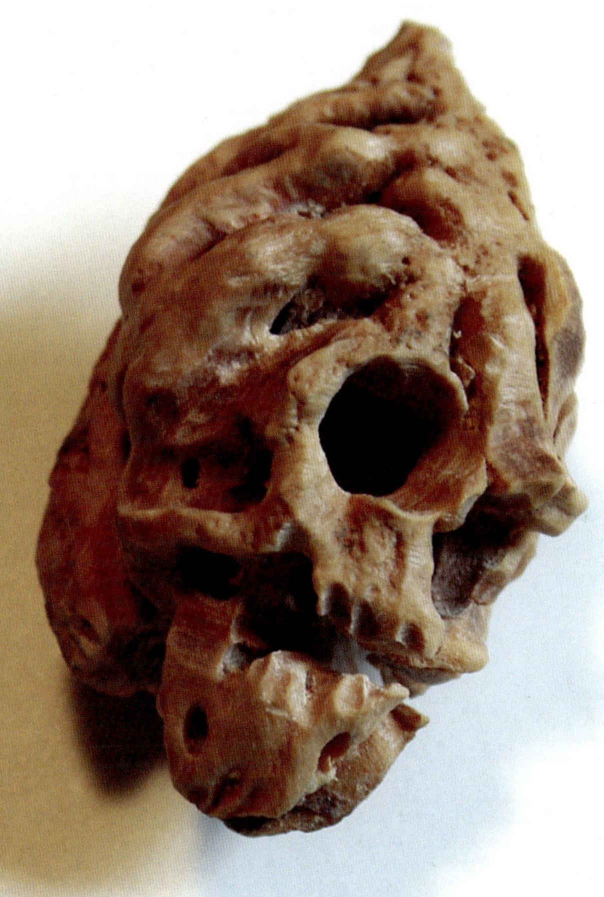

Peach Pit Skull Carved peach pit. (My grandfather used to carve peach pits, so this was done in his honor.)

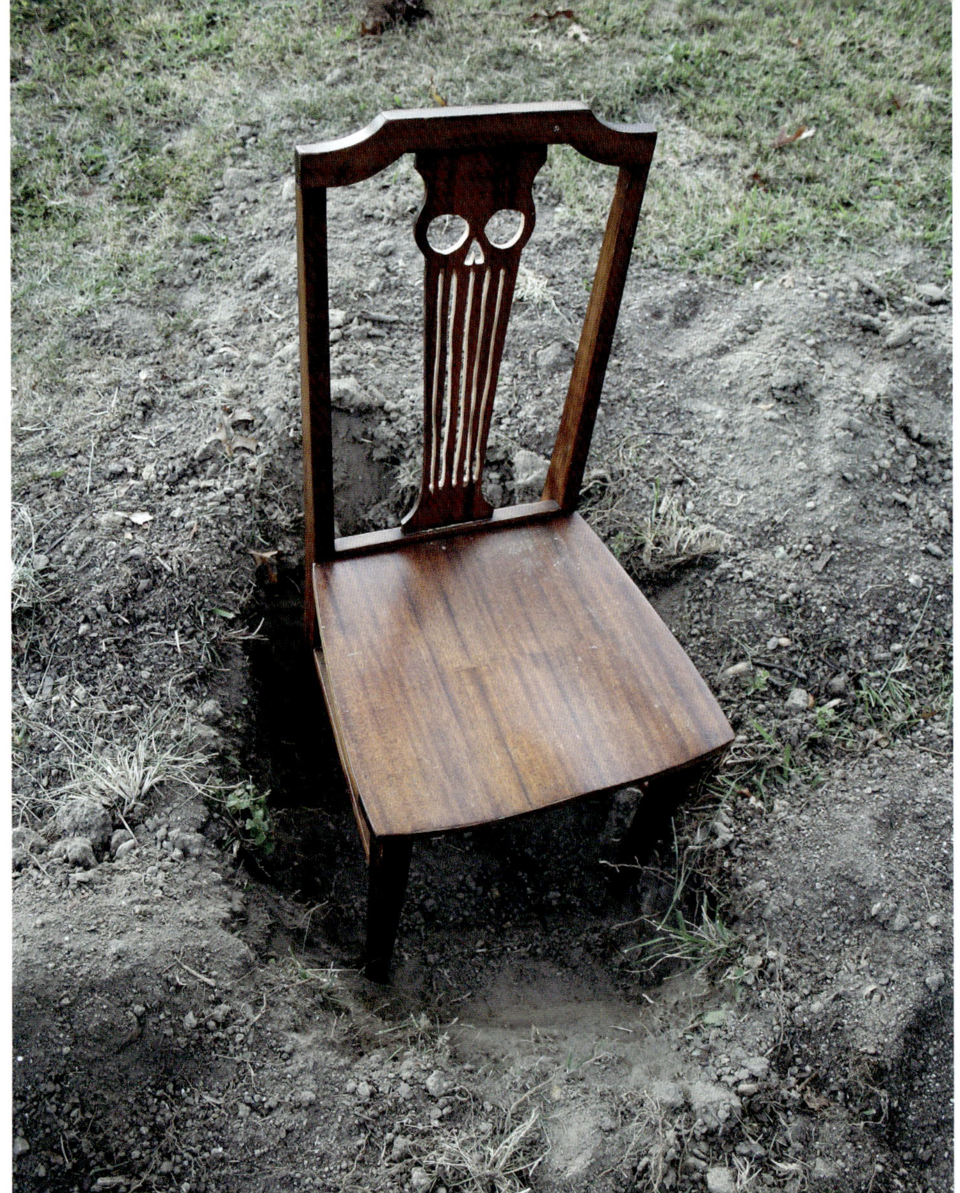

Skull Chair
Modified chair. (There was a hole in my yard the day I made this, so clearly I had to stick my chair in it.)

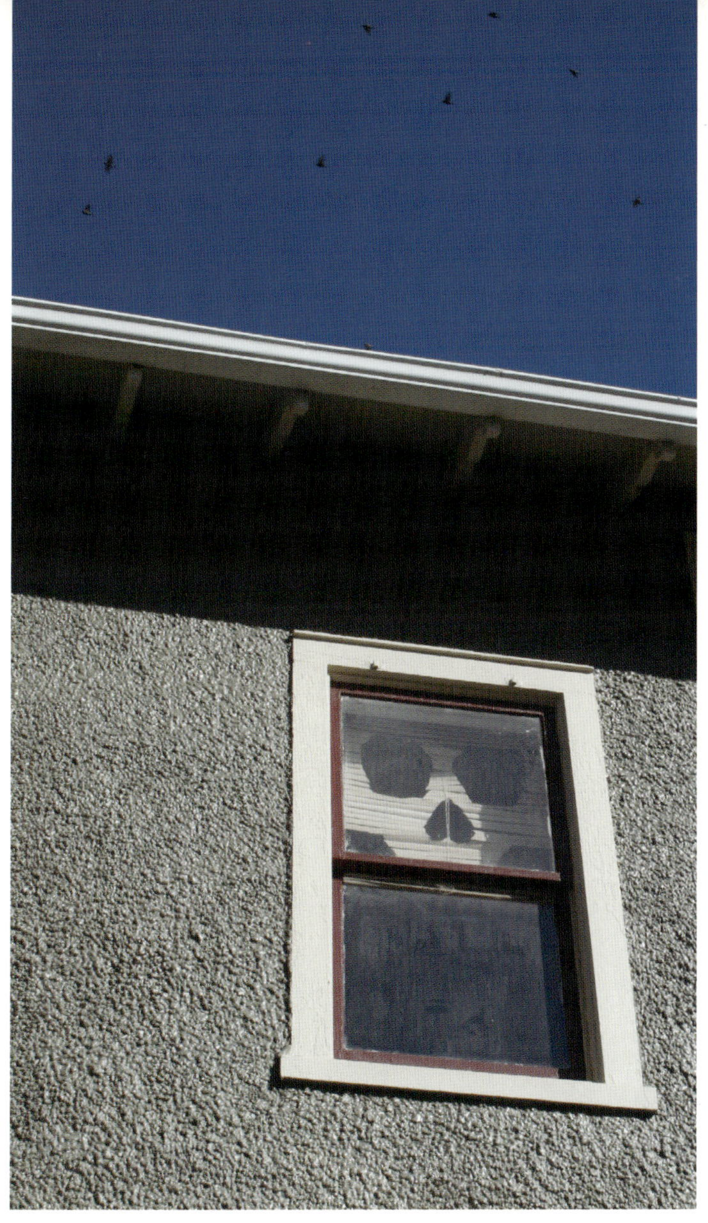

Shade(s of Death) Skull
Cut plastic window shade. (I've left this up since I made it, and the neighbors haven't complained yet!)

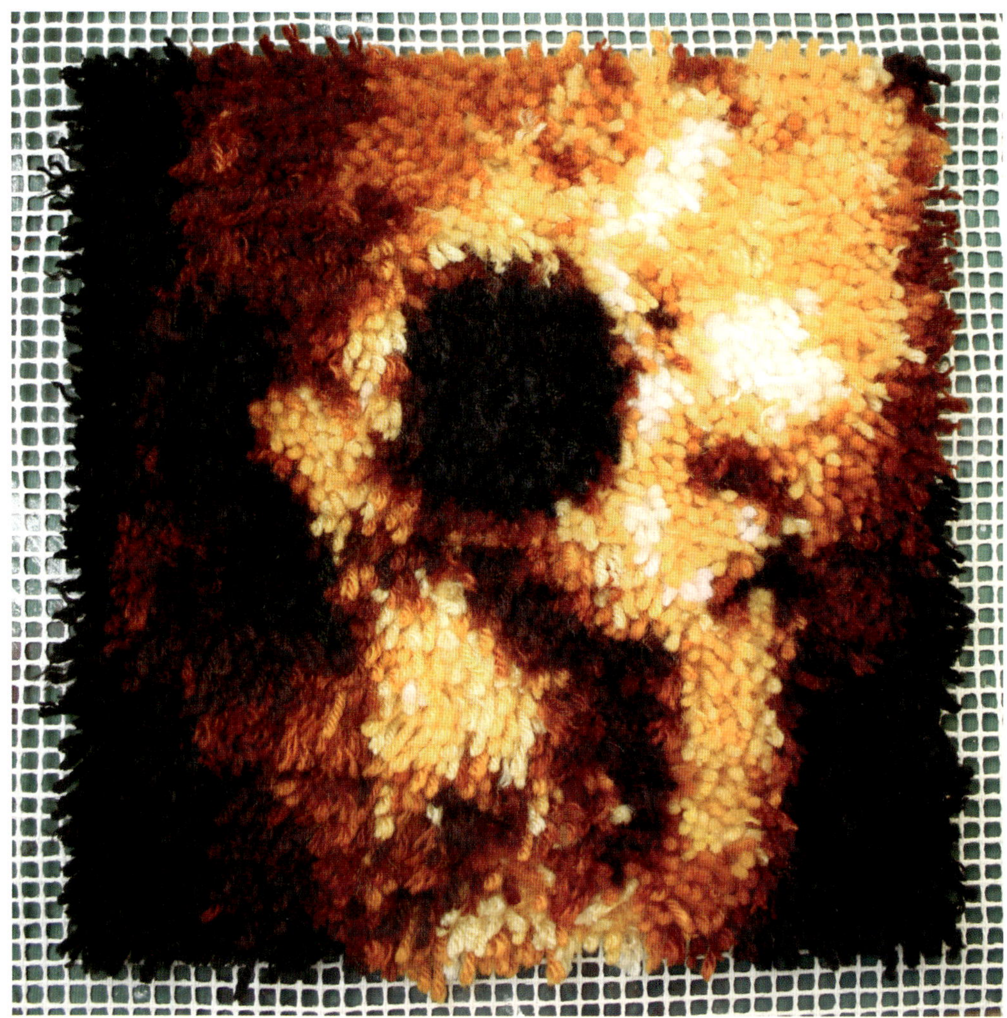

Latch Hook Skull Latch hooked yarn 11¾ x 11¾ inches (30 x 30cm). (I bought a pre-existing latch-hook kit of a puppy and made my own pattern using the yarn colors provided. This took nine hours to complete and gave me a migraine headache.)

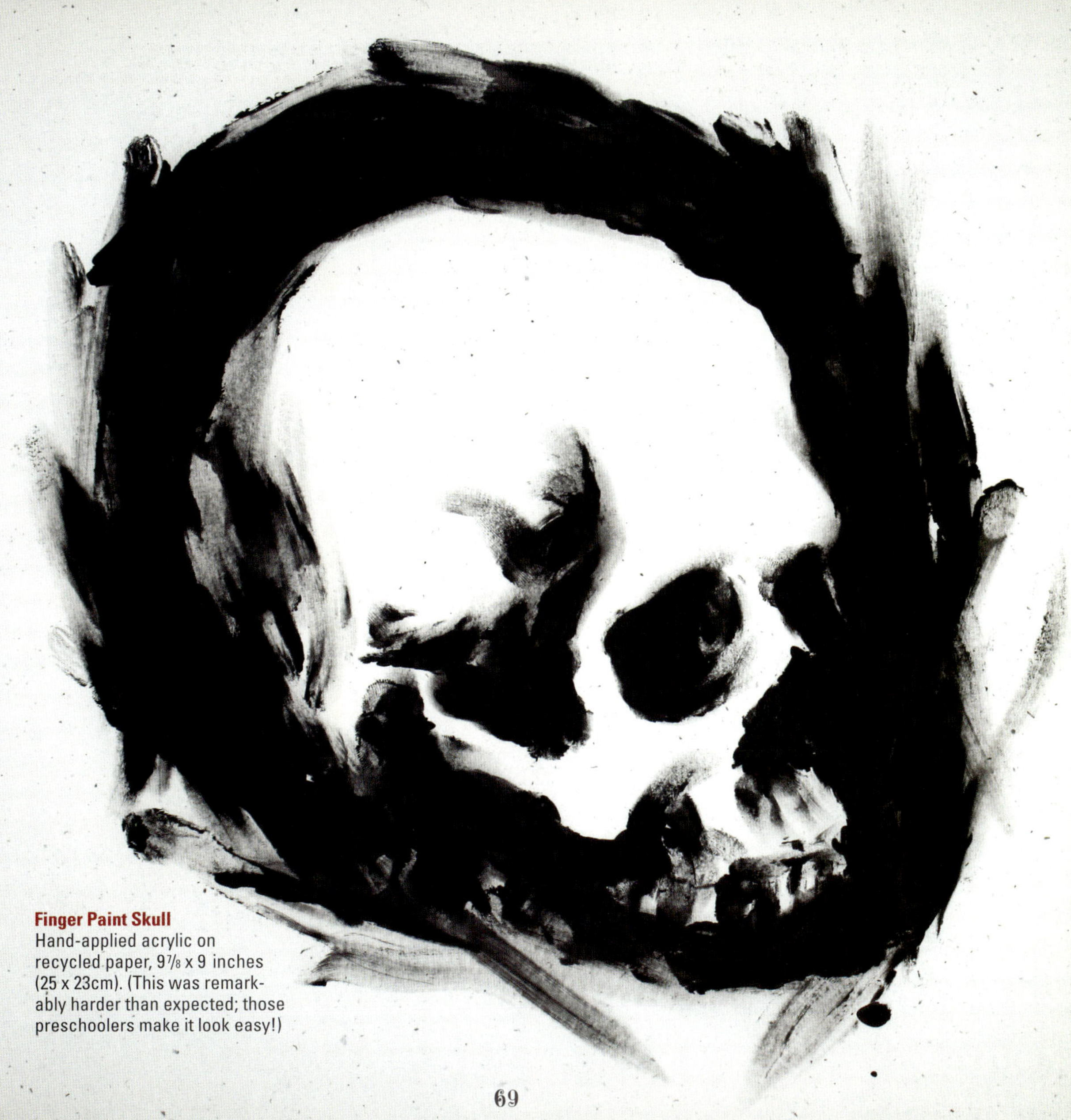

Finger Paint Skull
Hand-applied acrylic on recycled paper, 9⅞ x 9 inches (25 x 23cm). (This was remarkably harder than expected; those preschoolers make it look easy!)

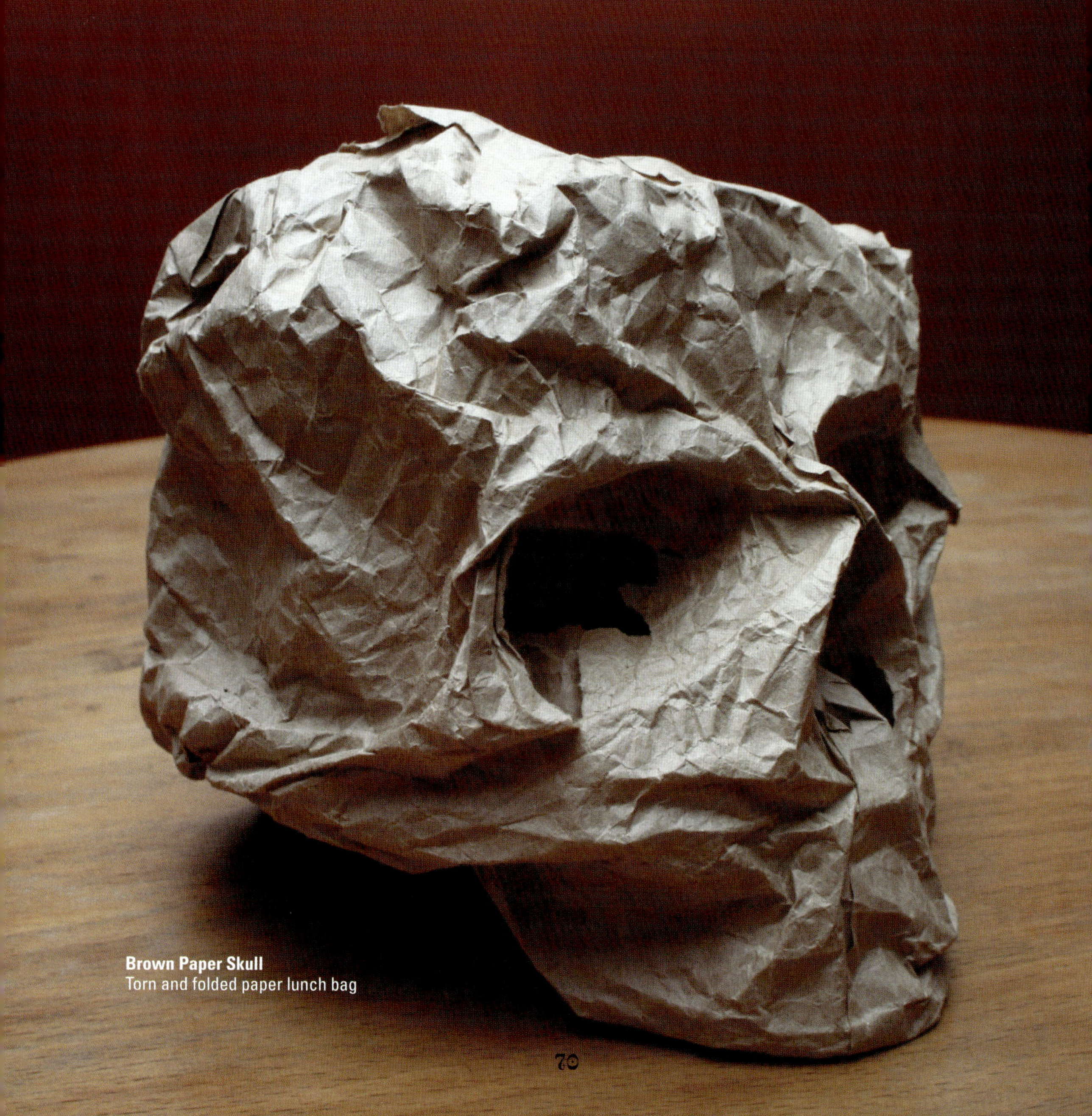

Brown Paper Skull
Torn and folded paper lunch bag

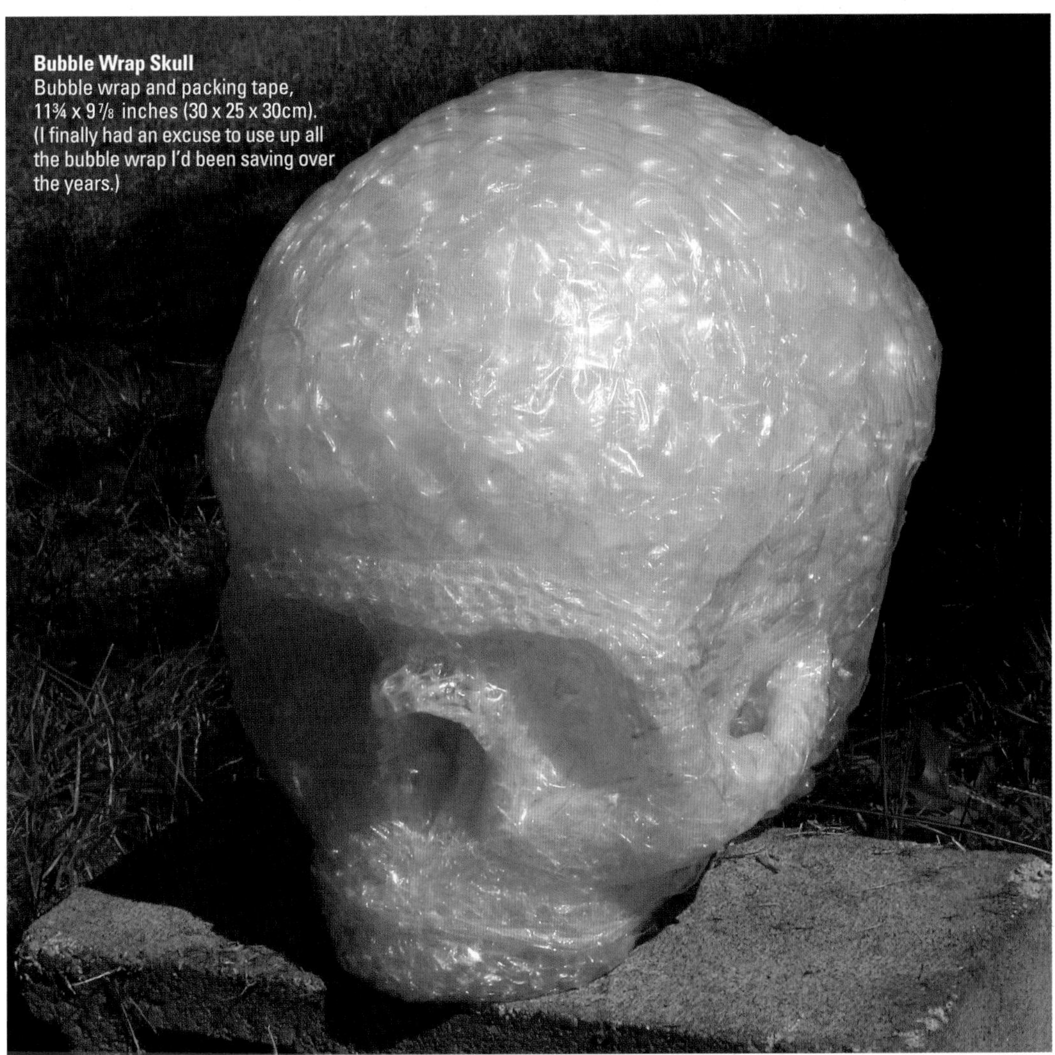

Bubble Wrap Skull
Bubble wrap and packing tape, 11¾ x 9⅞ inches (30 x 25 x 30cm). (I finally had an excuse to use up all the bubble wrap I'd been saving over the years.)

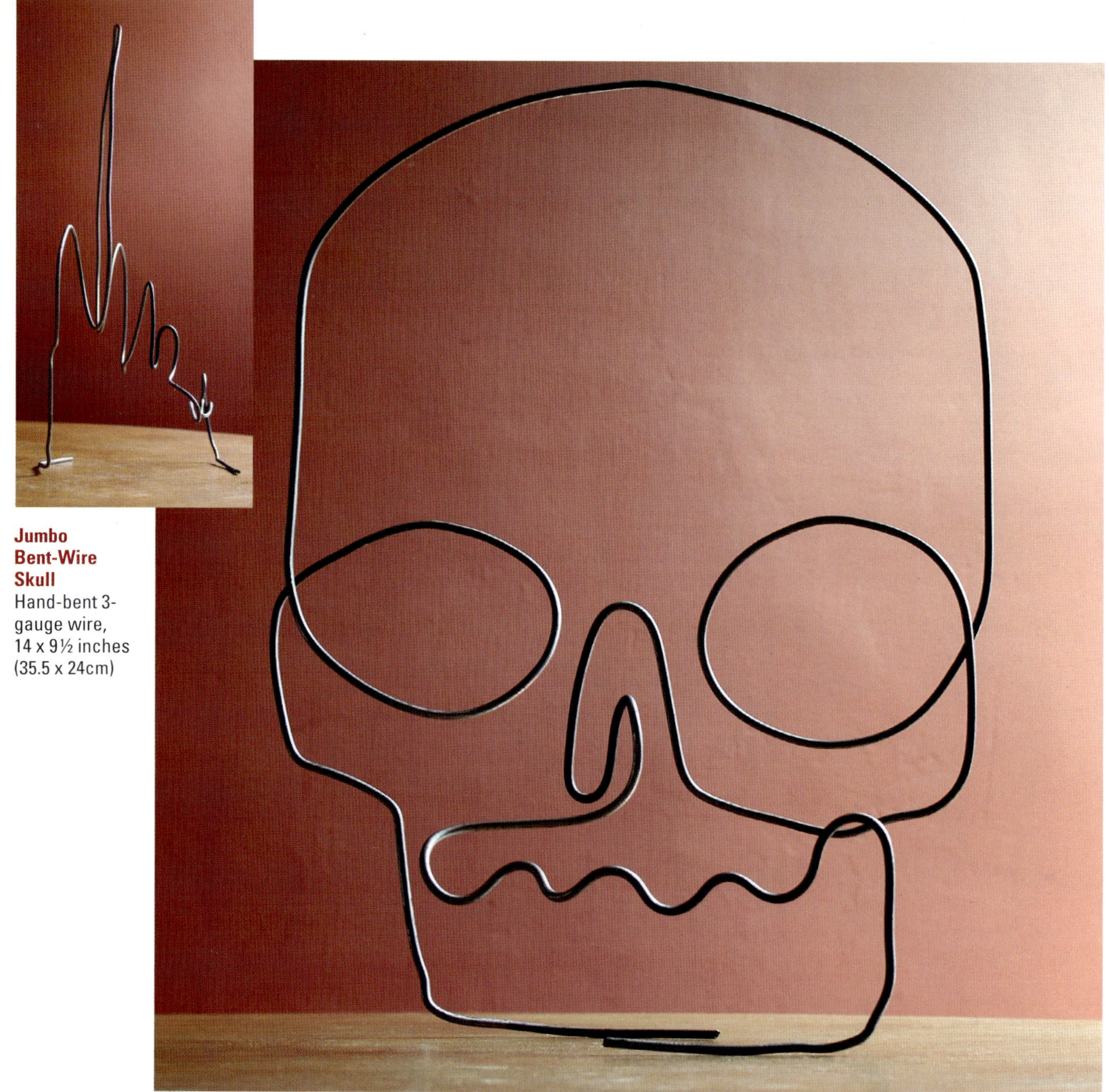

Jumbo Bent-Wire Skull
Hand-bent 3-gauge wire, 14 x 9½ inches (35.5 x 24cm)

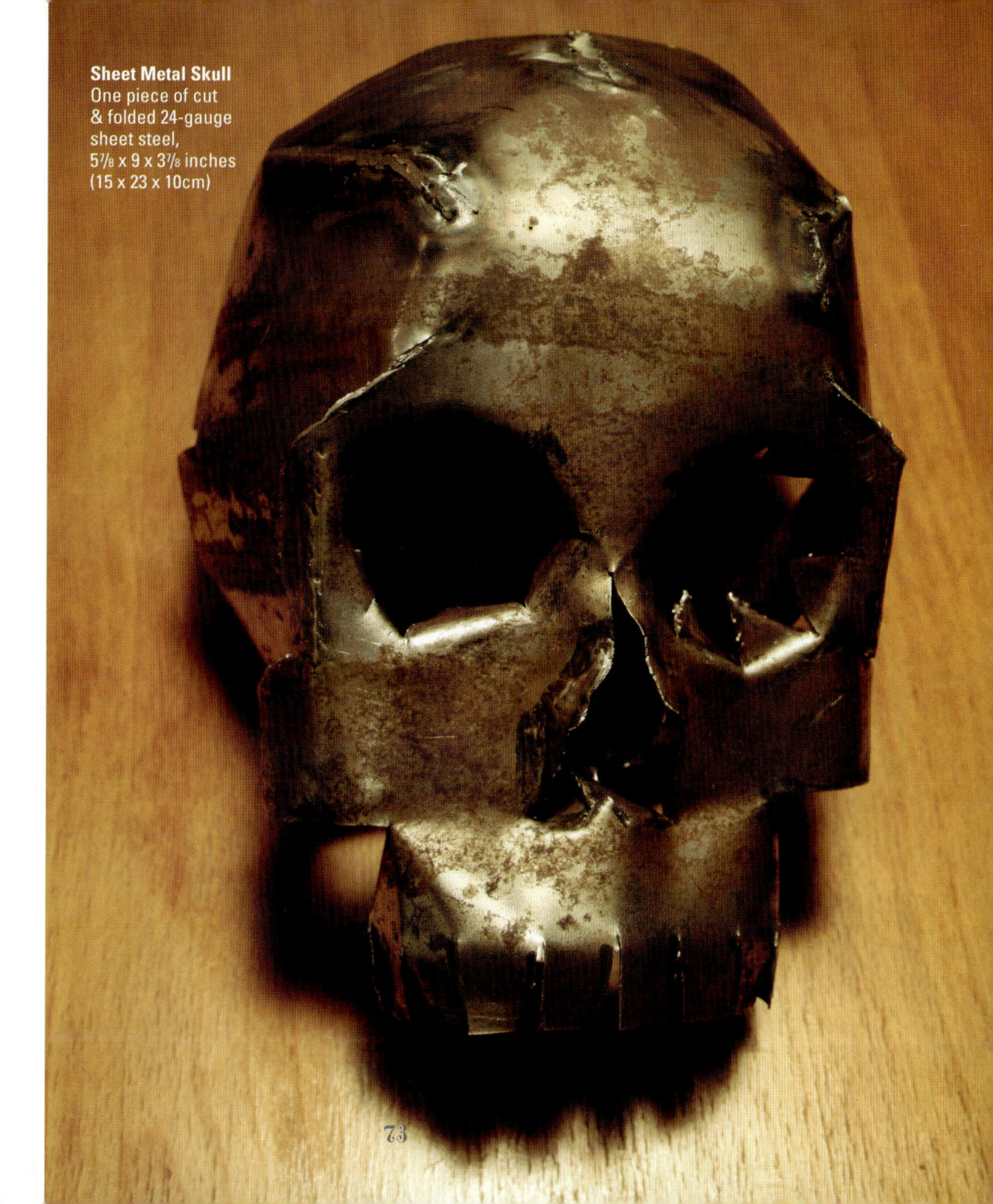

Sheet Metal Skull
One piece of cut & folded 24-gauge sheet steel, $5\frac{7}{8} \times 9 \times 3\frac{7}{8}$ inches (15 x 23 x 10cm)

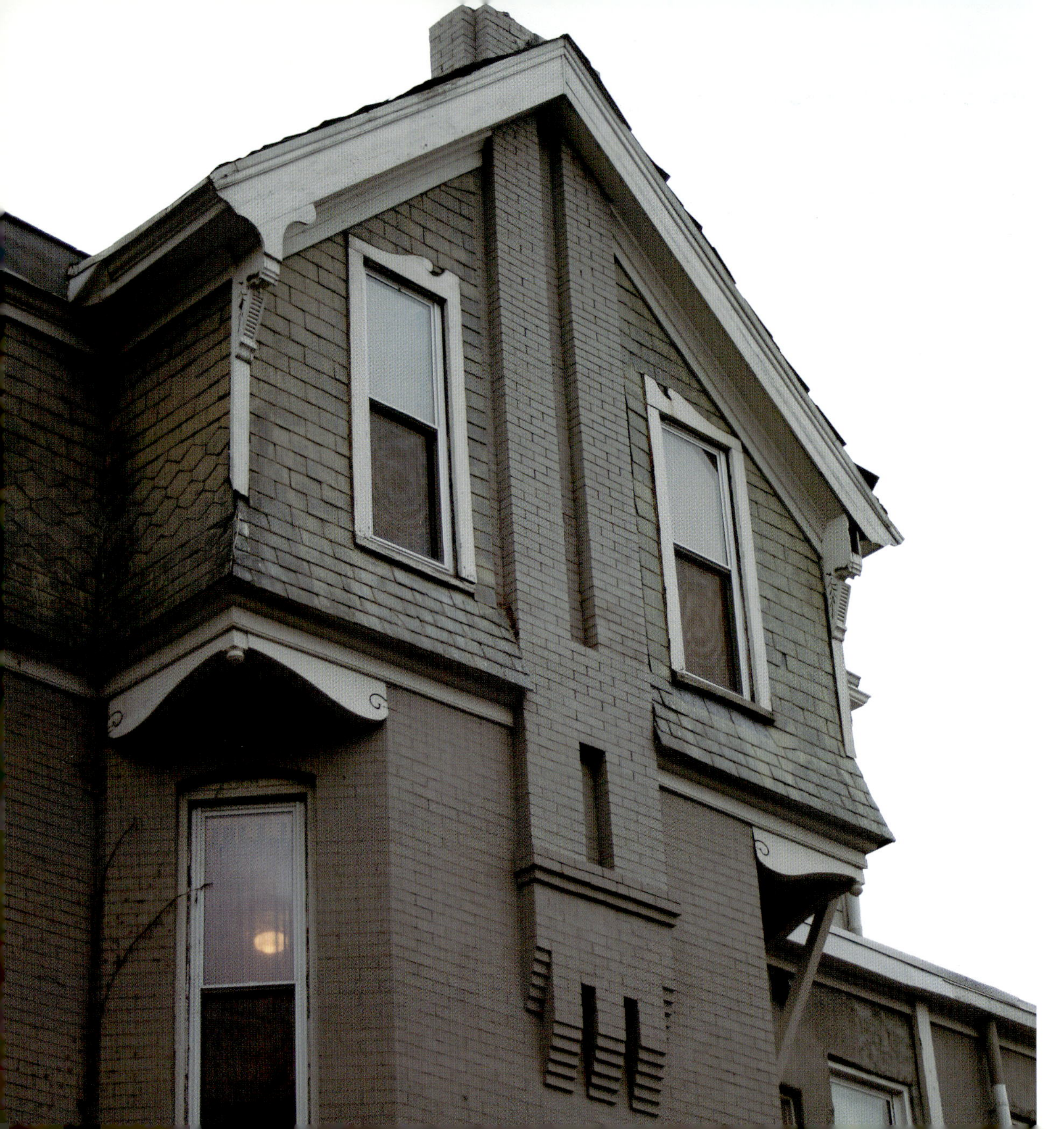

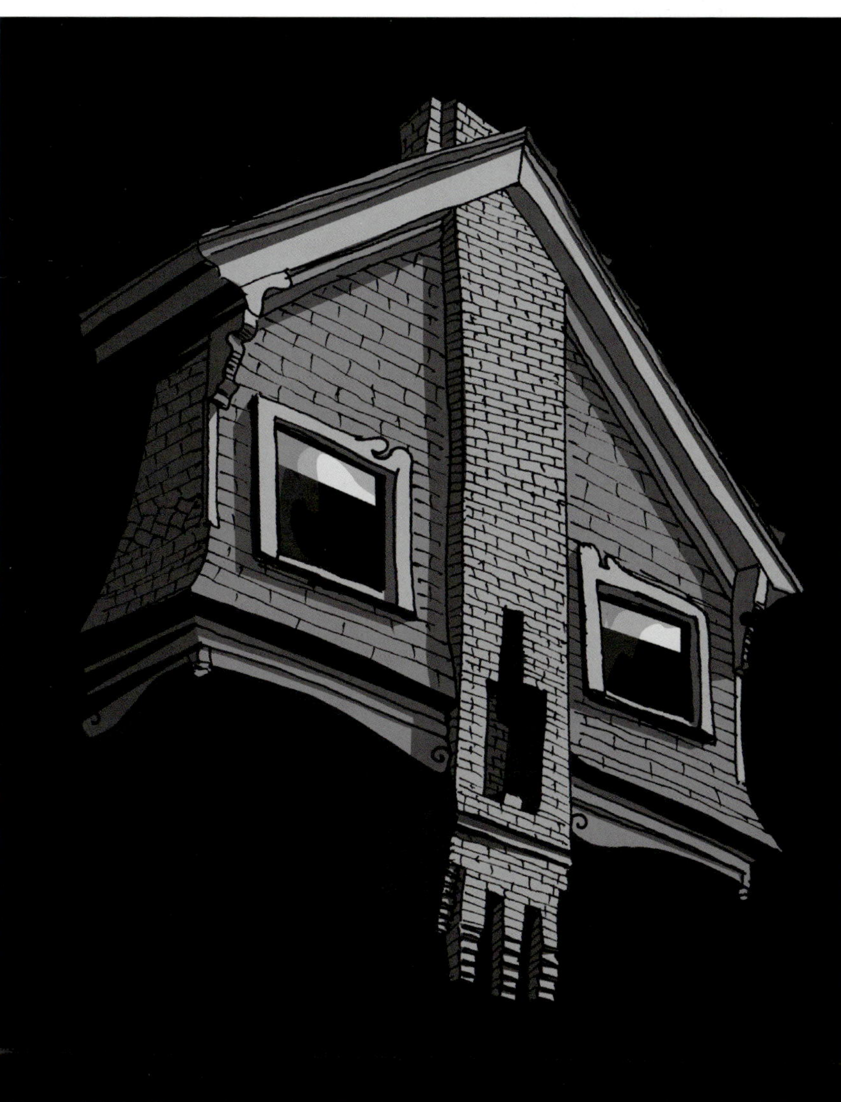

Apartment Grayskull
Photography and digitally colored ink pen illustration. (My friends actually used to live in this apartment and gave it the name Grayskull.)

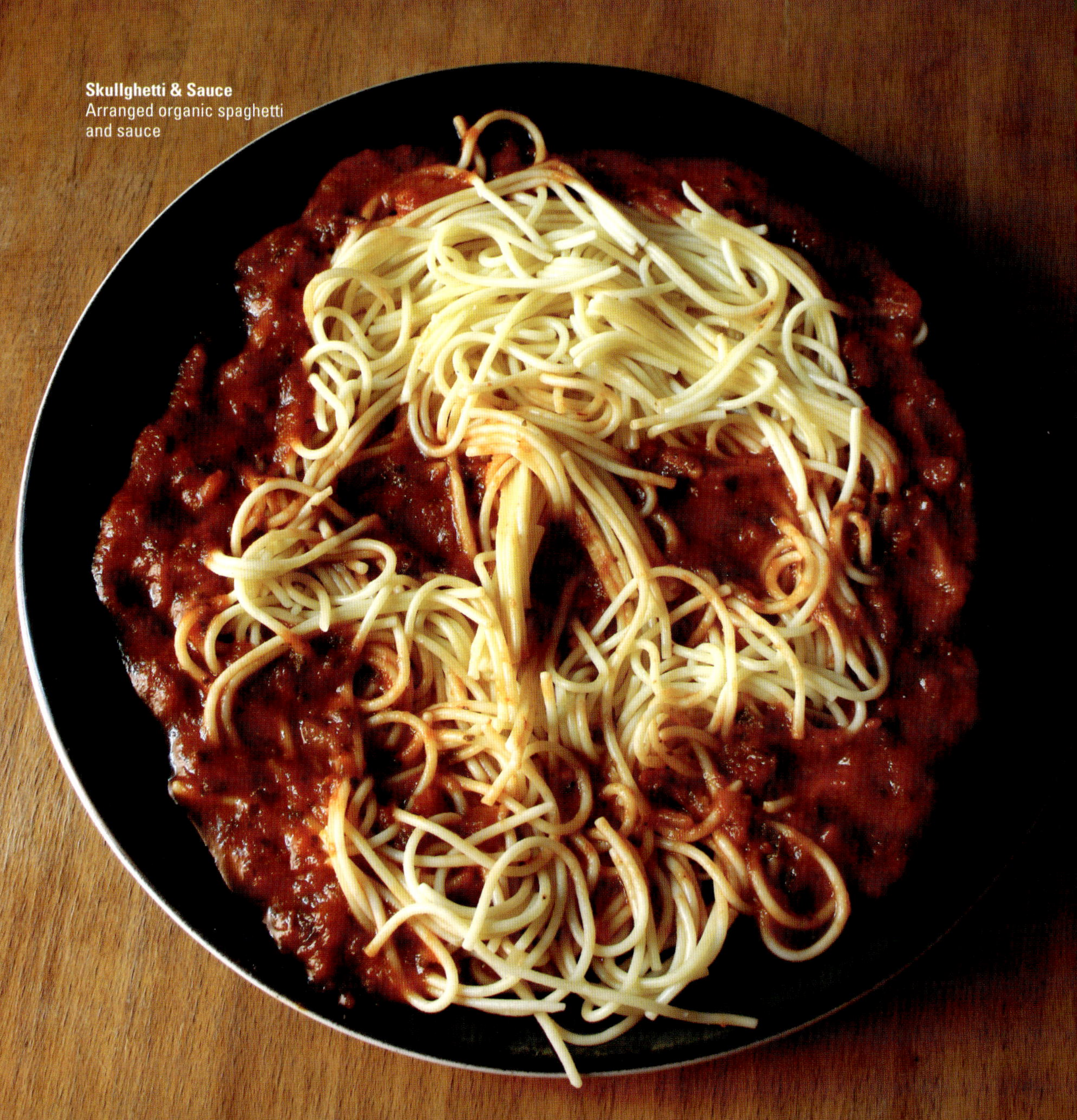

Skullghetti & Sauce
Arranged organic spaghetti and sauce

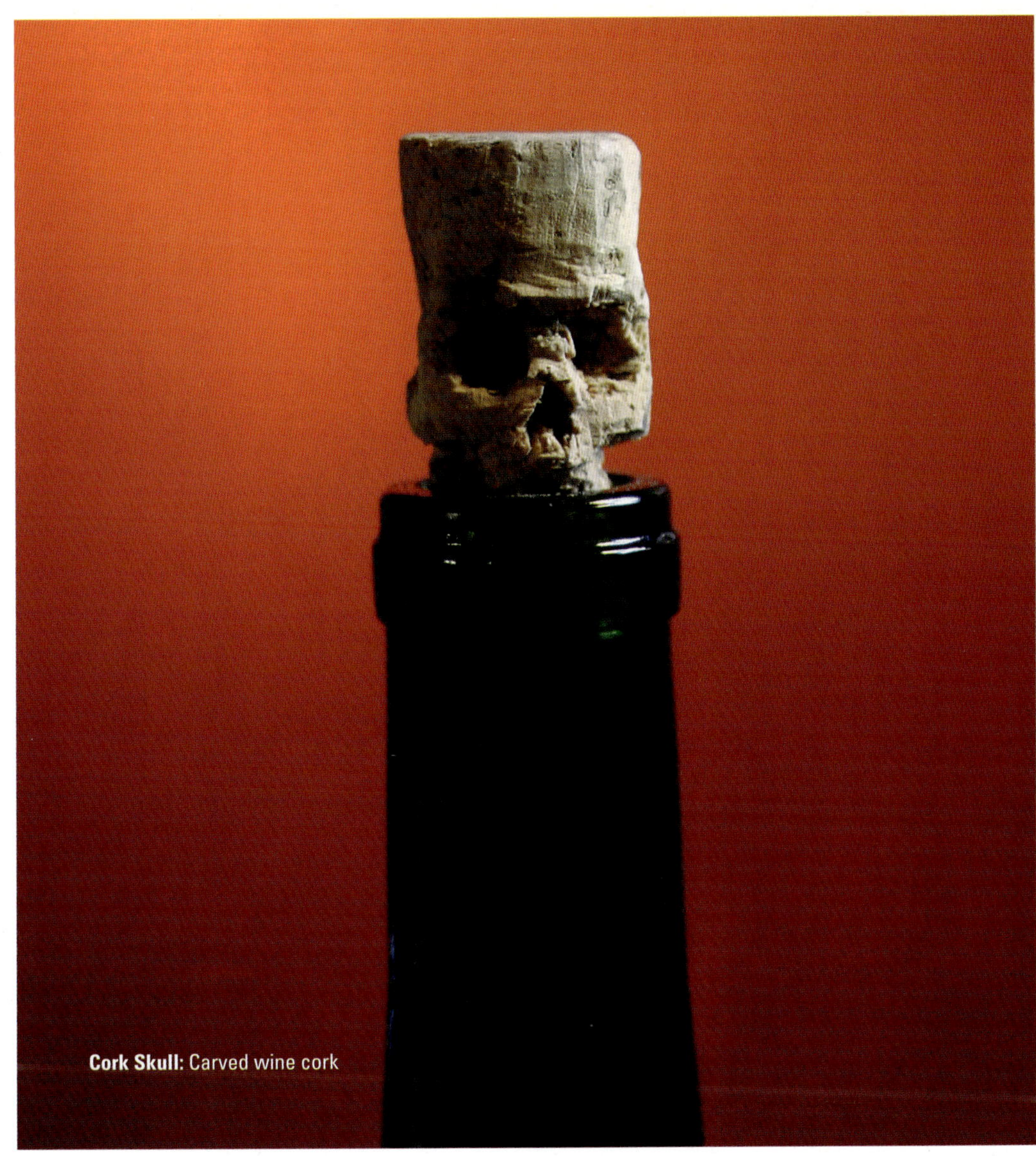

Cork Skull: Carved wine cork

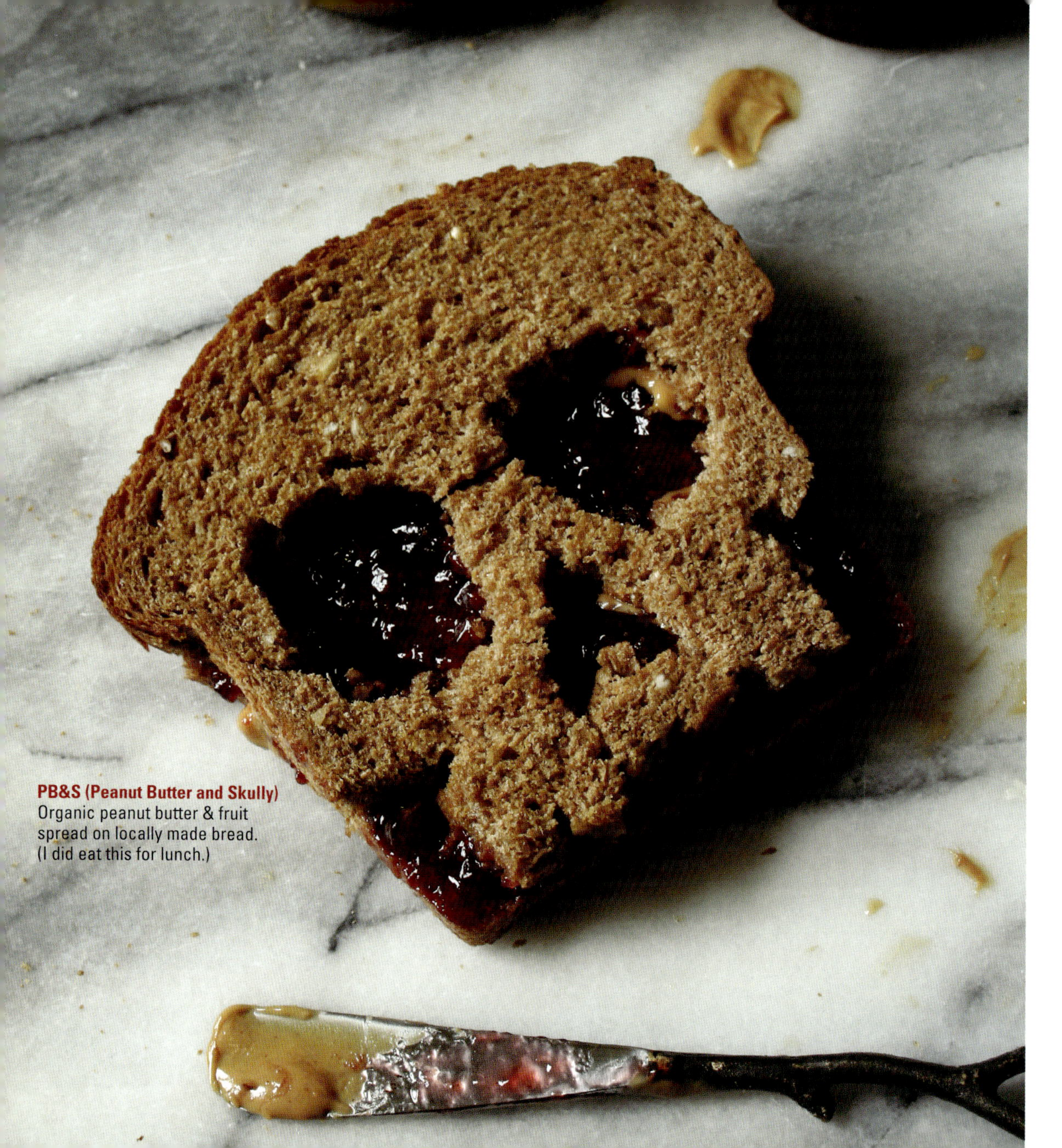

PB&S (Peanut Butter and Skully)
Organic peanut butter & fruit spread on locally made bread. (I did eat this for lunch.)

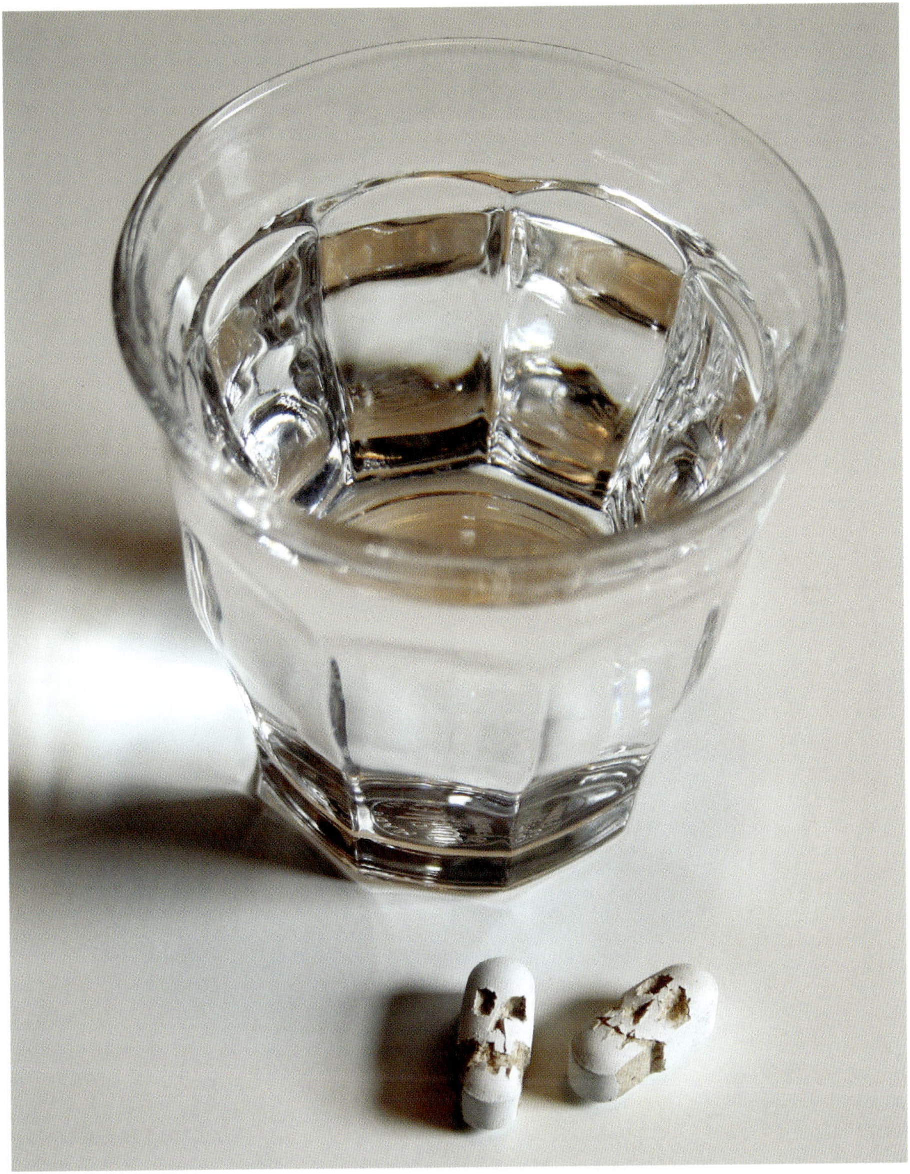

Take Two Skulls
Carved vitamin supplement pills. (I took these from my dad. Luckily he lived without them.)

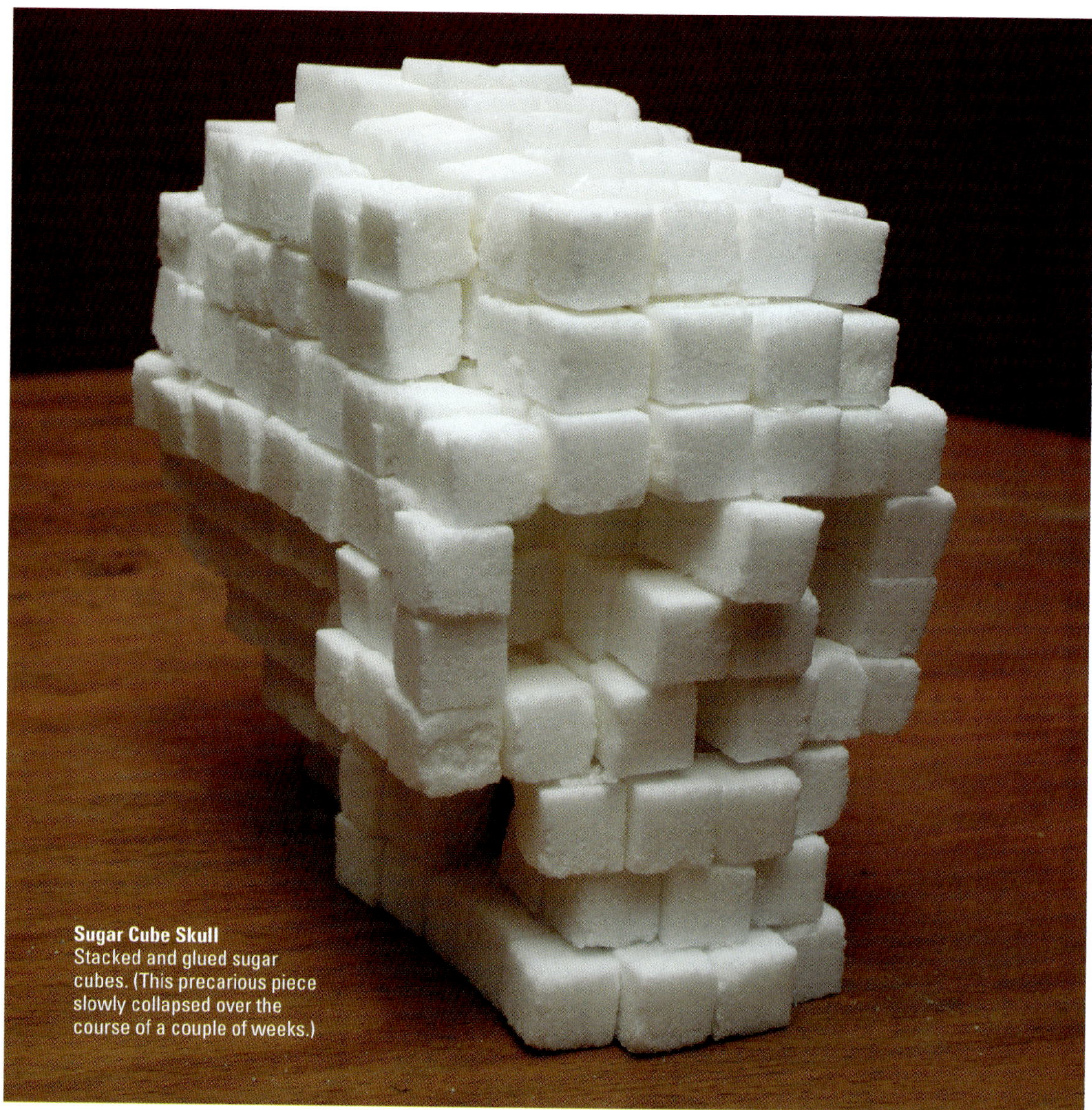

Sugar Cube Skull
Stacked and glued sugar cubes. (This precarious piece slowly collapsed over the course of a couple of weeks.)

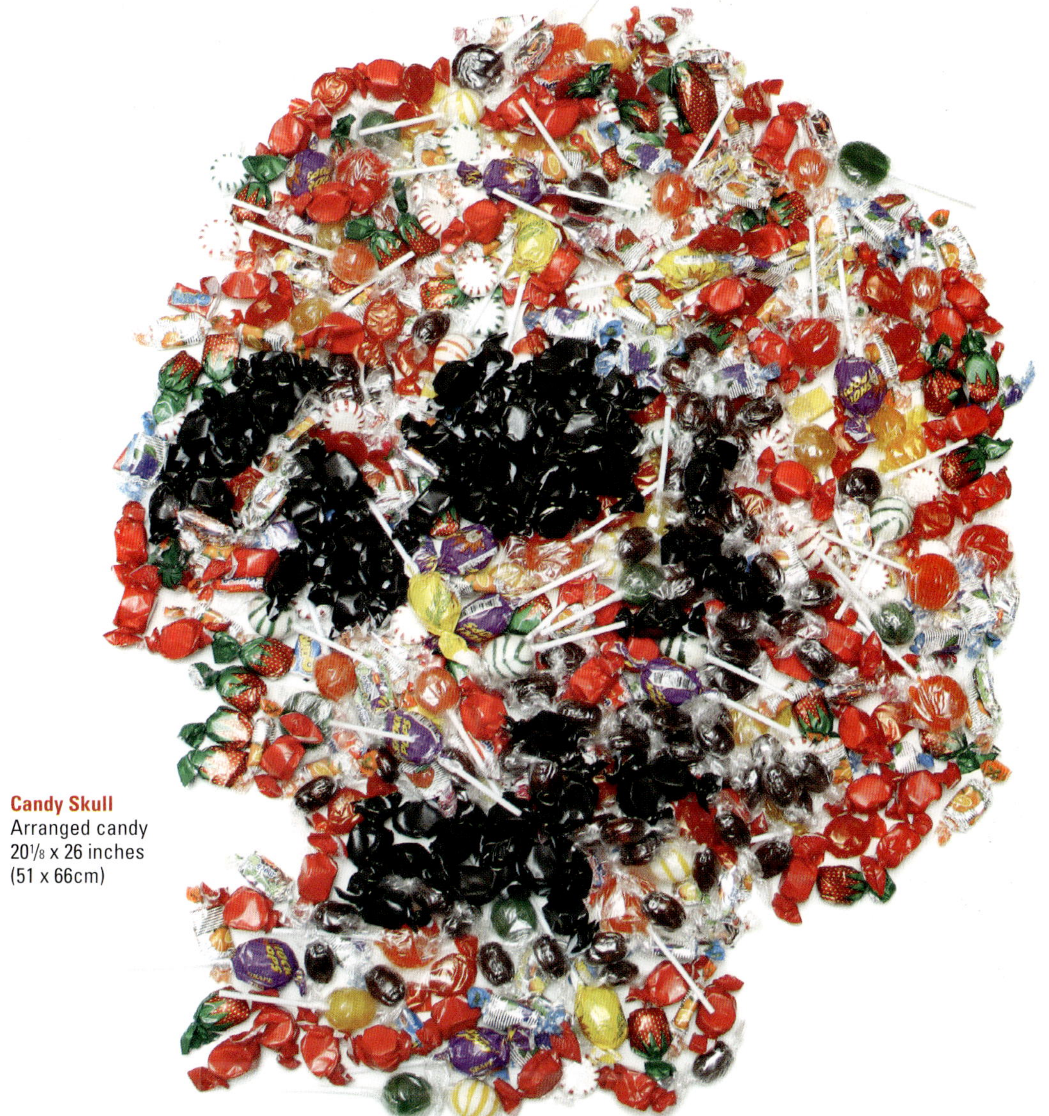

Candy Skull
Arranged candy
20⅛ x 26 inches
(51 x 66cm)

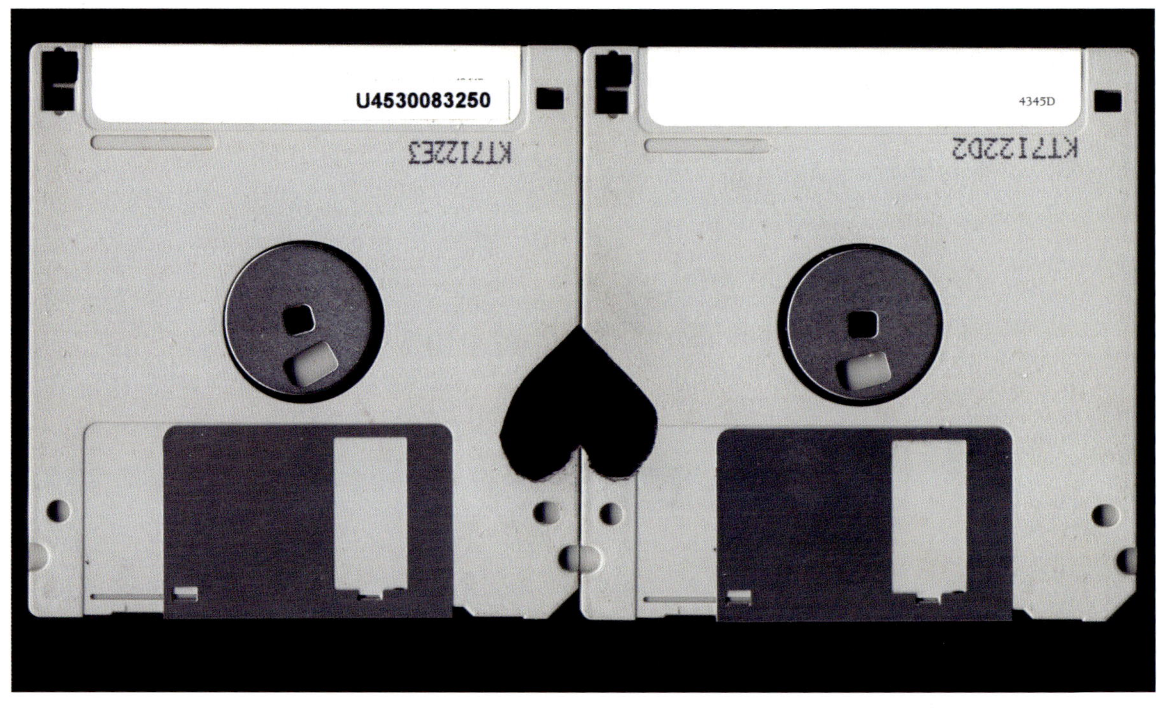

Floppy Skull 1.0
Cut 3½ inch floppy discs

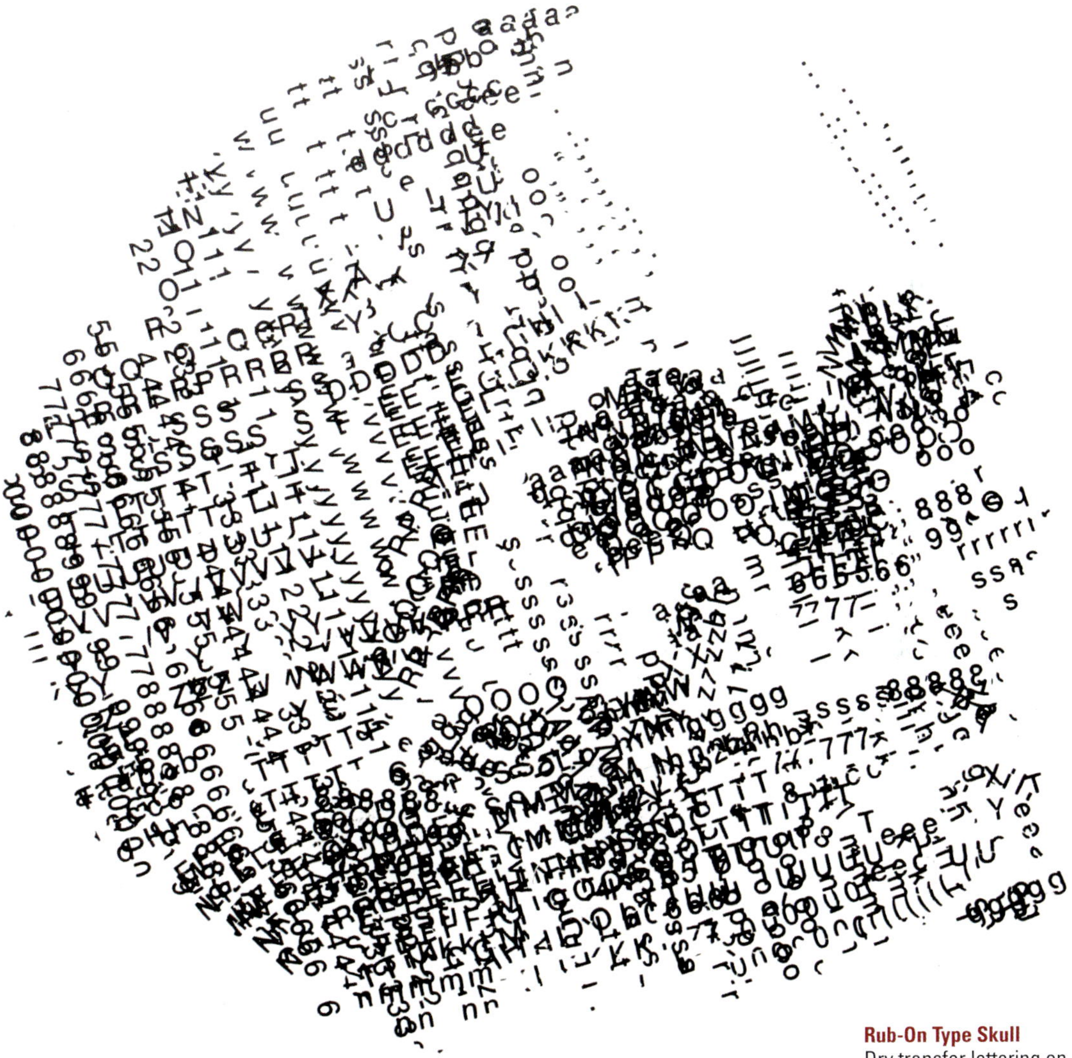

Rub-On Type Skull

Dry transfer lettering on board, 5 1/8 x 5 1/8 inches (13 x 13cm)

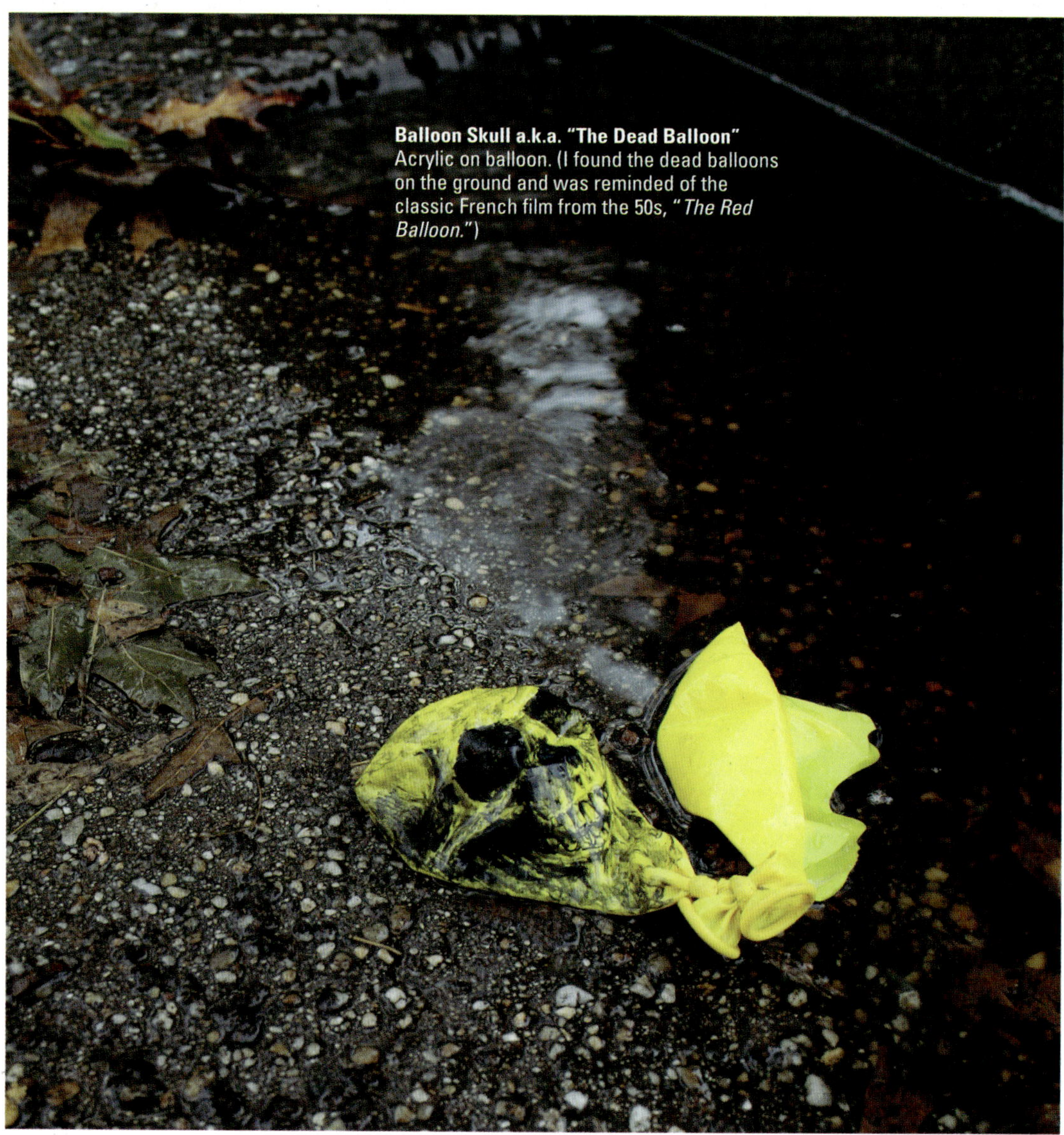

Balloon Skull a.k.a. "The Dead Balloon"
Acrylic on balloon. (I found the dead balloons on the ground and was reminded of the classic French film from the 50s, "*The Red Balloon.*")

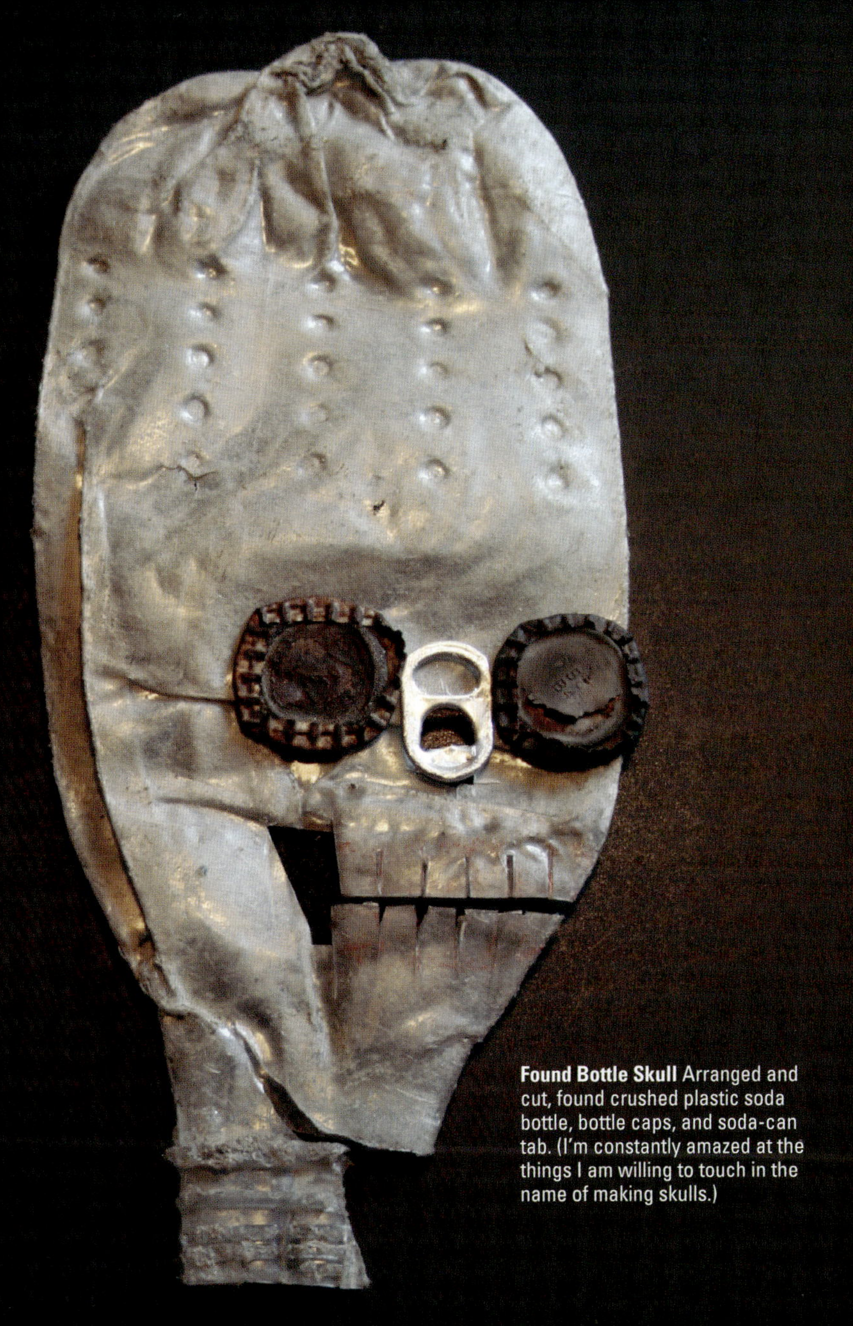

Found Bottle Skull Arranged and cut, found crushed plastic soda bottle, bottle caps, and soda-can tab. (I'm constantly amazed at the things I am willing to touch in the name of making skulls.)

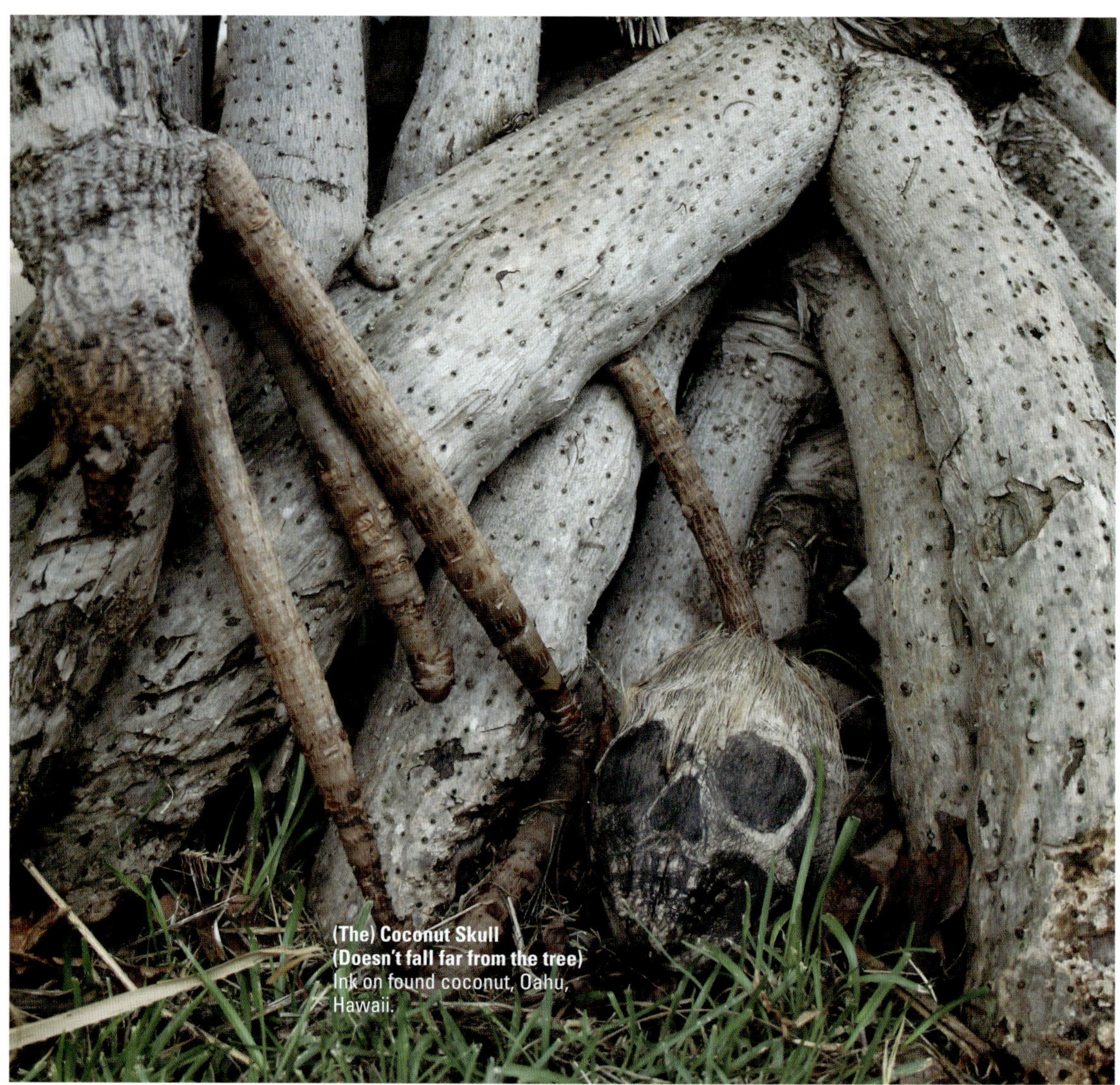

(The) Coconut Skull
(Doesn't fall far from the tree)
Ink on found coconut, Oahu, Hawaii.

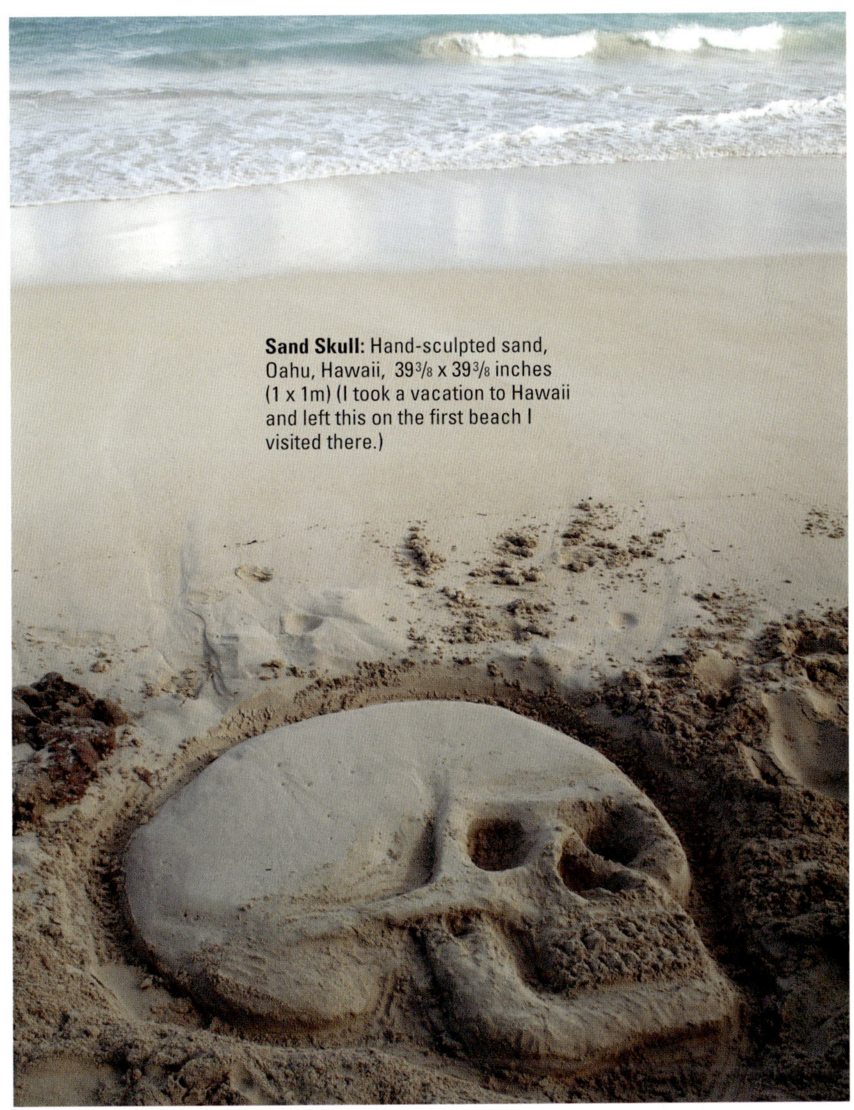

Sand Skull: Hand-sculpted sand, Oahu, Hawaii, 39³/₈ x 39³/₈ inches (1 x 1m) (I took a vacation to Hawaii and left this on the first beach I visited there.)

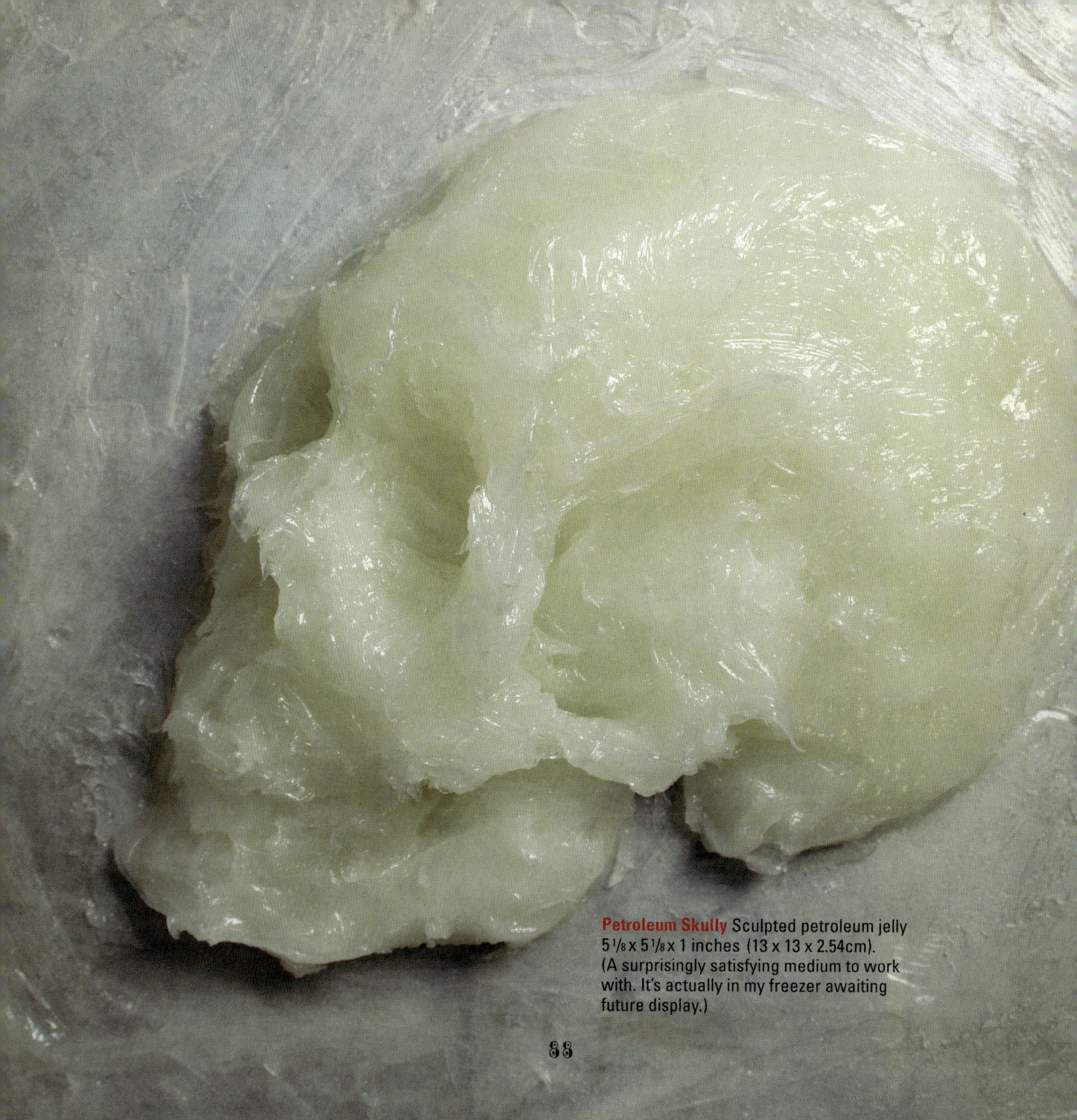

Petroleum Skully Sculpted petroleum jelly 5 1/8 x 5 1/8 x 1 inches (13 x 13 x 2.54cm). (A surprisingly satisfying medium to work with. It's actually in my freezer awaiting future display.)

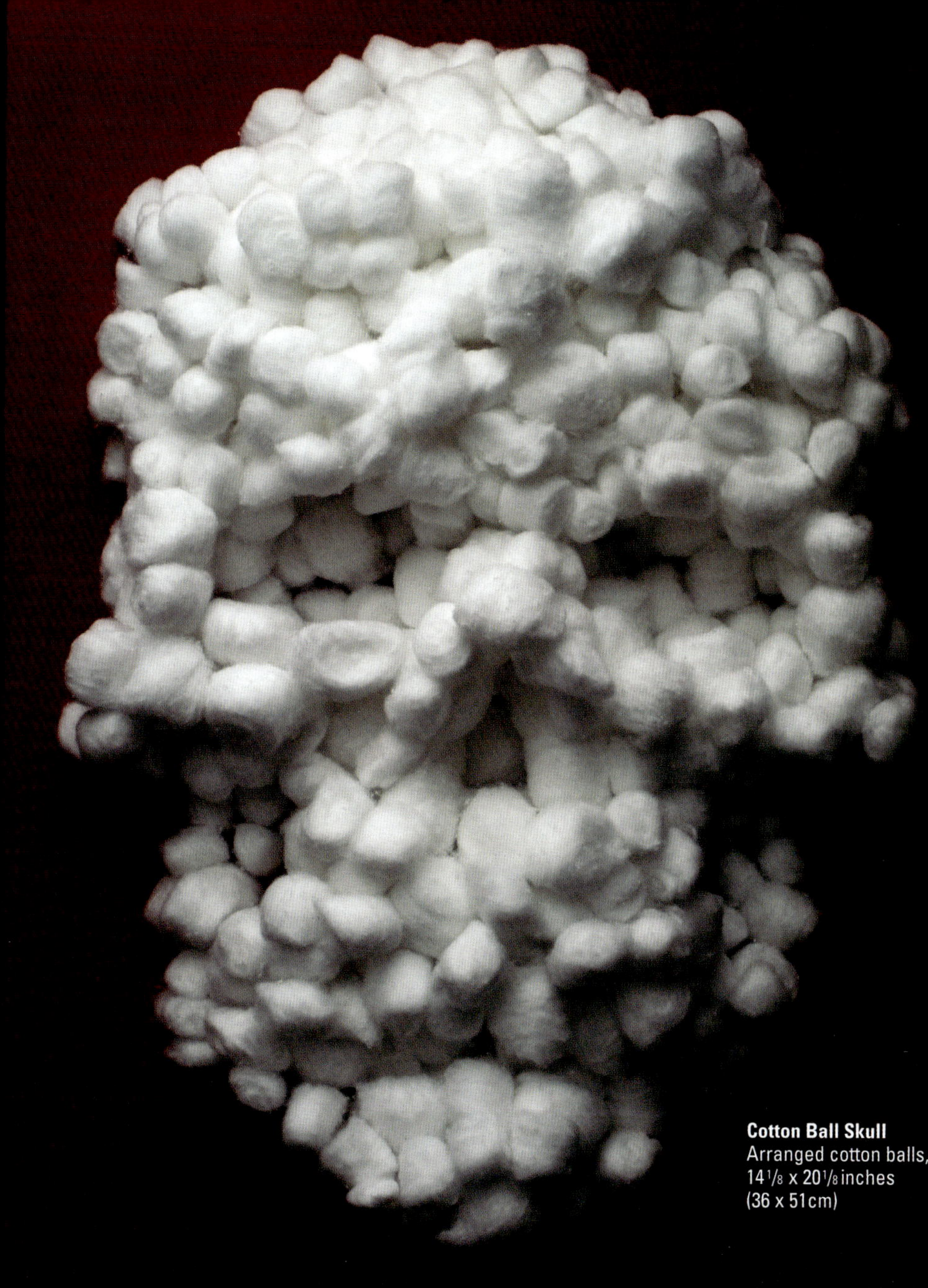

Cotton Ball Skull
Arranged cotton balls,
14 1/8 x 20 1/8 inches
(36 x 51cm)

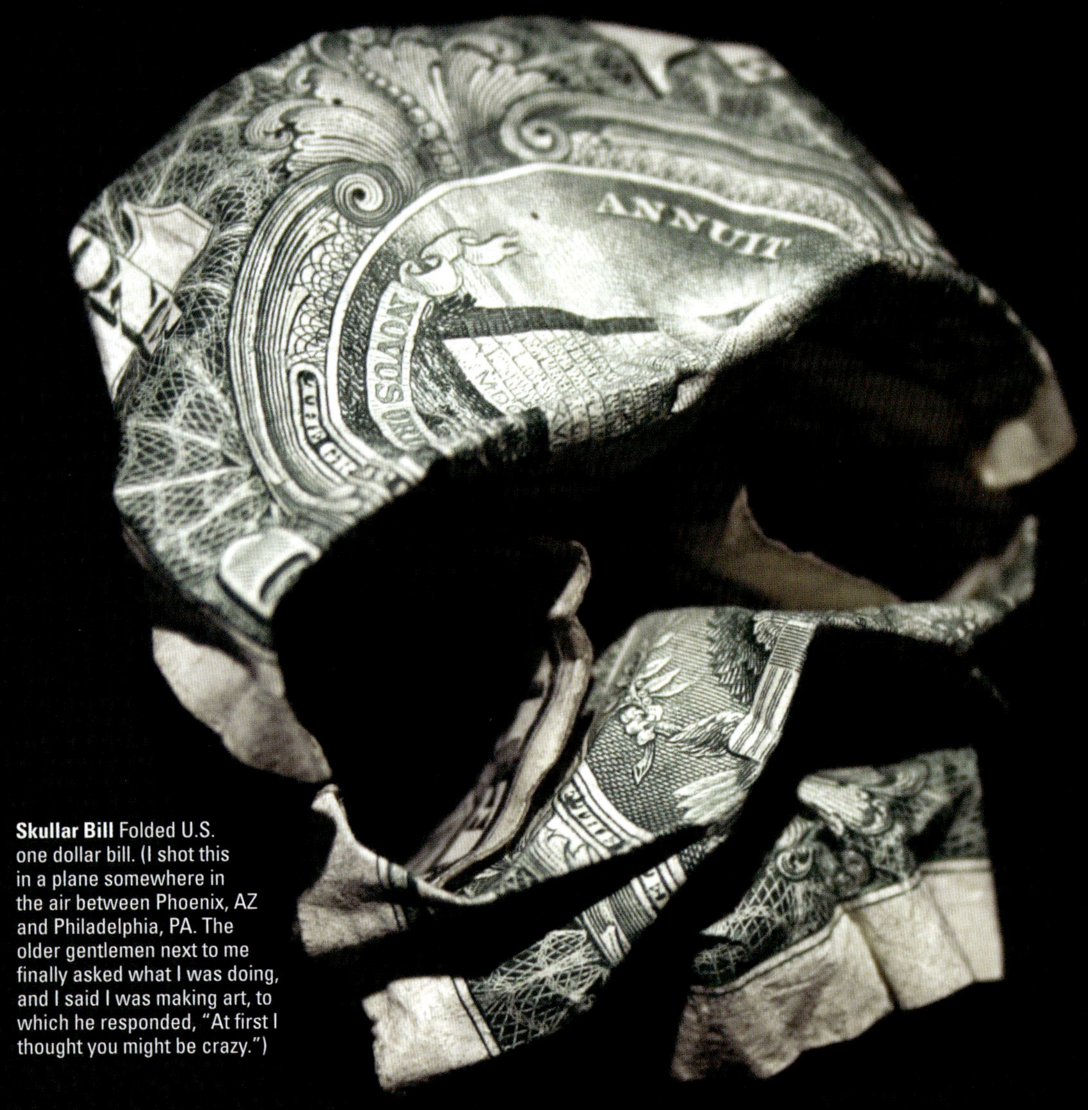

Skullar Bill Folded U.S. one dollar bill. (I shot this in a plane somewhere in the air between Phoenix, AZ and Philadelphia, PA. The older gentlemen next to me finally asked what I was doing, and I said I was making art, to which he responded, "At first I thought you might be crazy.")

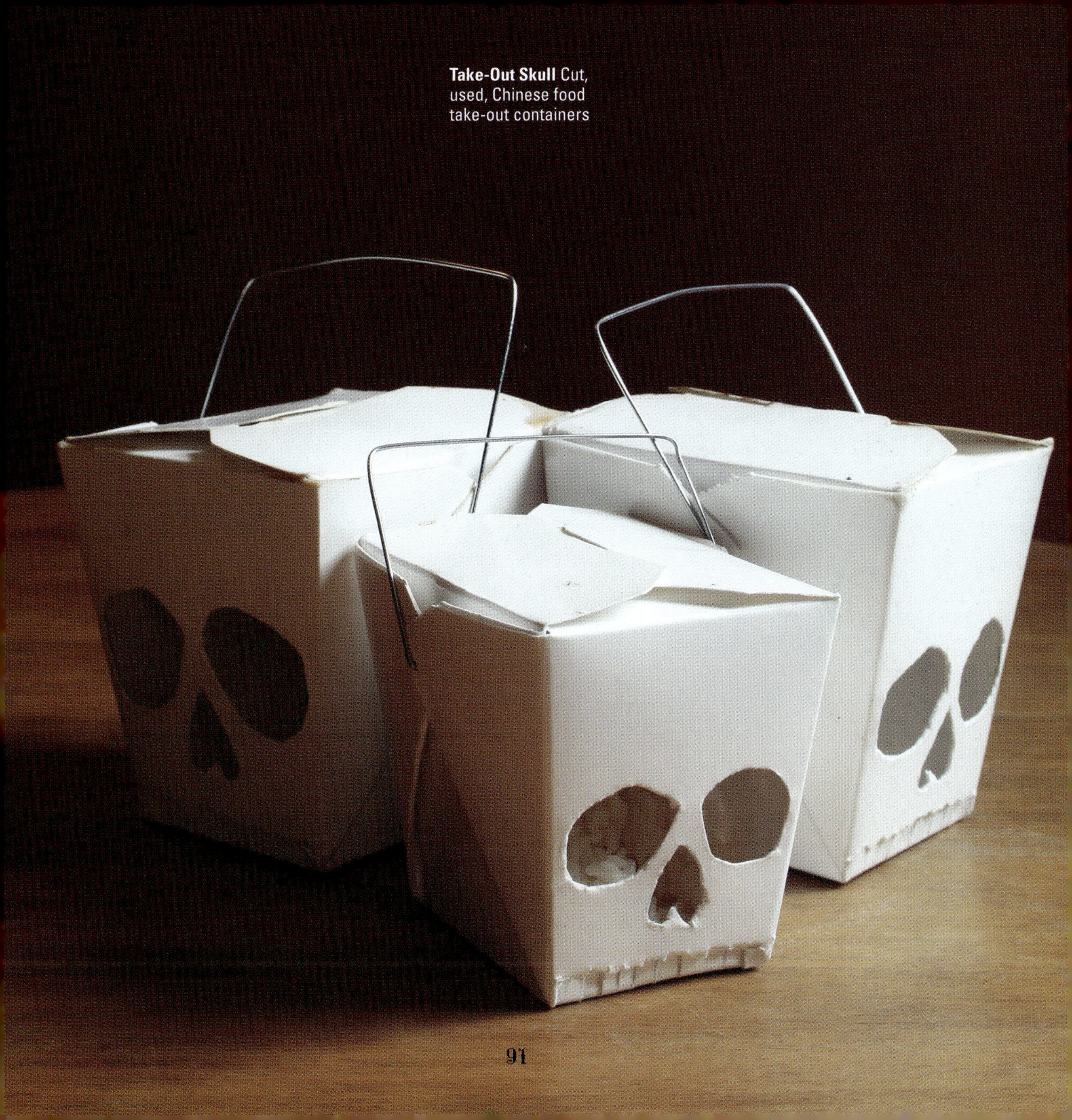

Take-Out Skull Cut, used, Chinese food take-out containers

Flower Skull Arranged flower petals
Previously published on the cover of *Poetry* Magazine,
December 2007

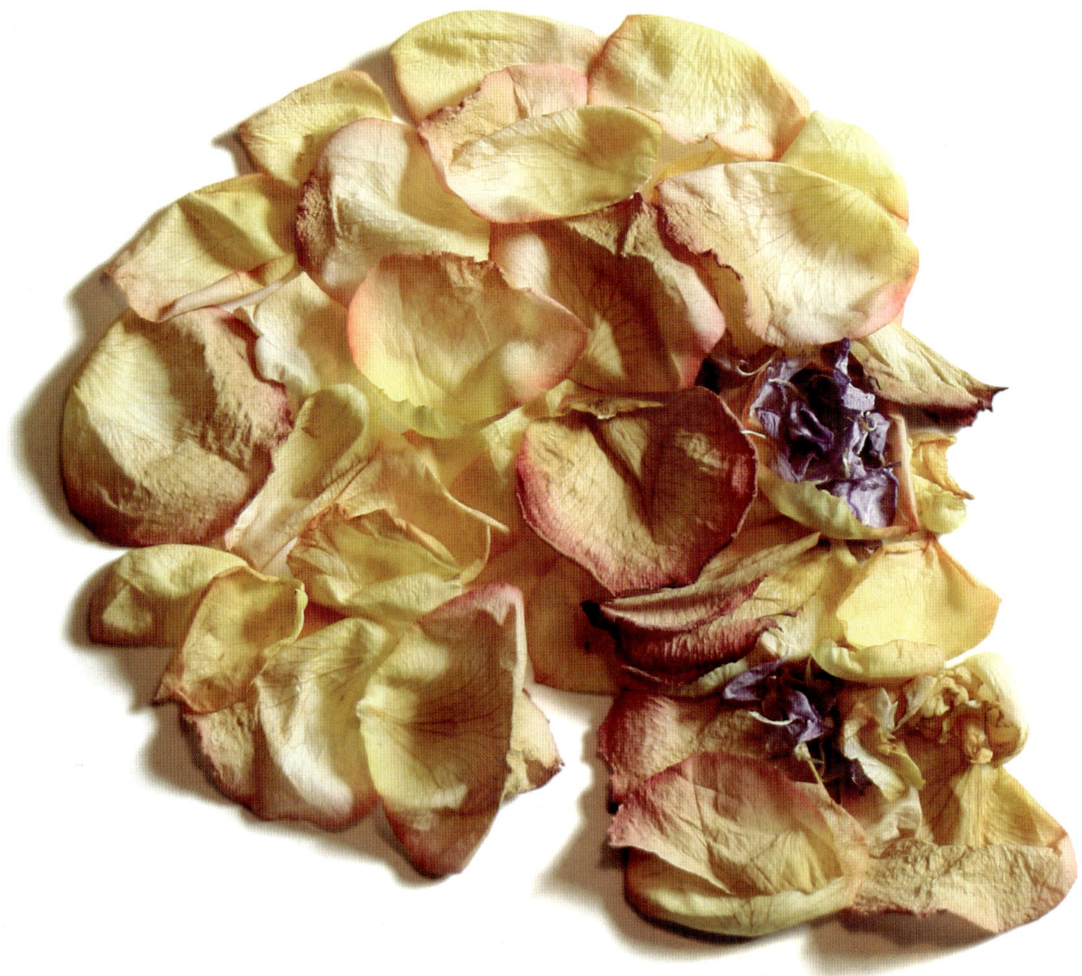

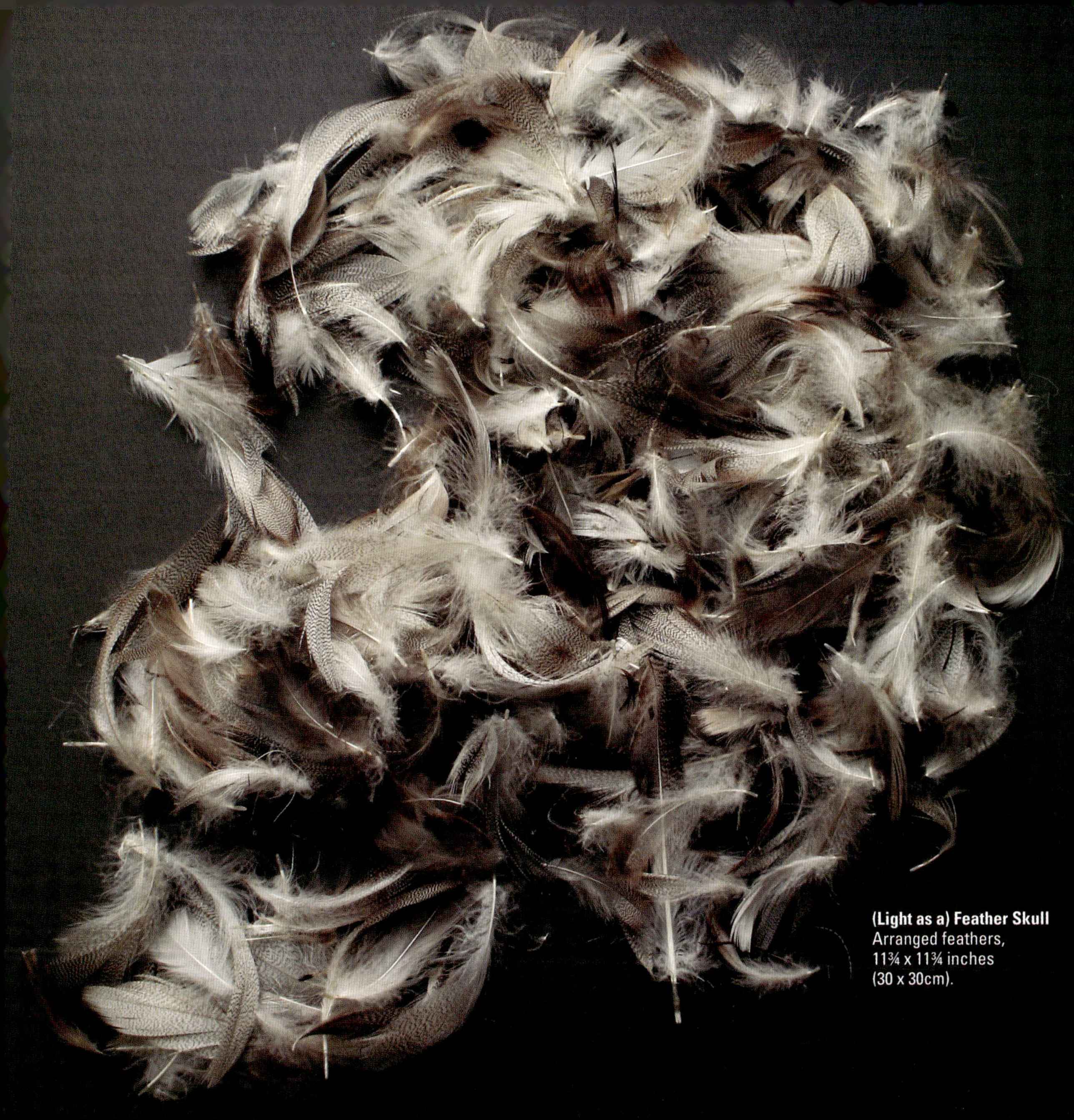

(Light as a) Feather Skull
Arranged feathers,
11¾ x 11¾ inches
(30 x 30cm).

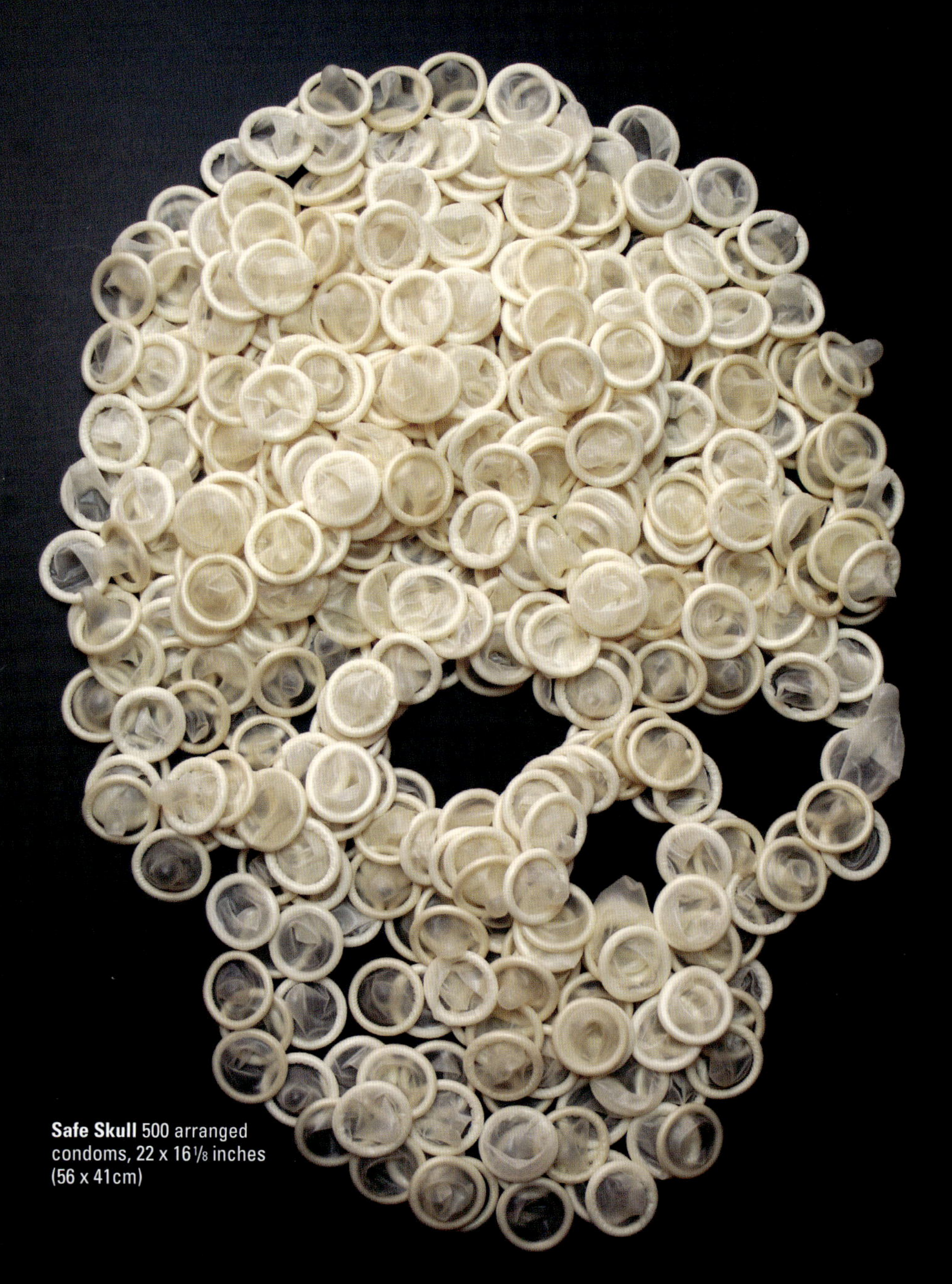

Safe Skull 500 arranged condoms, 22 x 16⅛ inches (56 x 41cm)

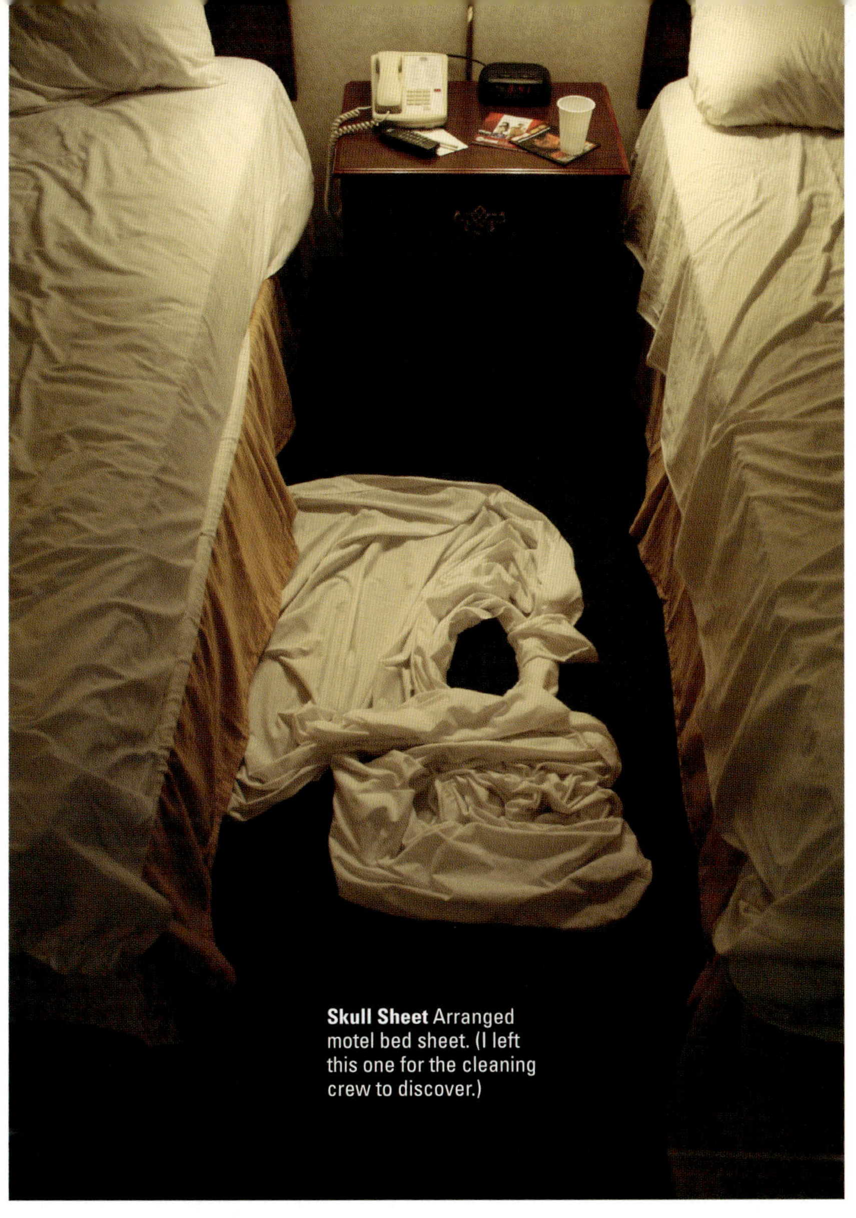

Skull Sheet Arranged motel bed sheet. (I left this one for the cleaning crew to discover.)

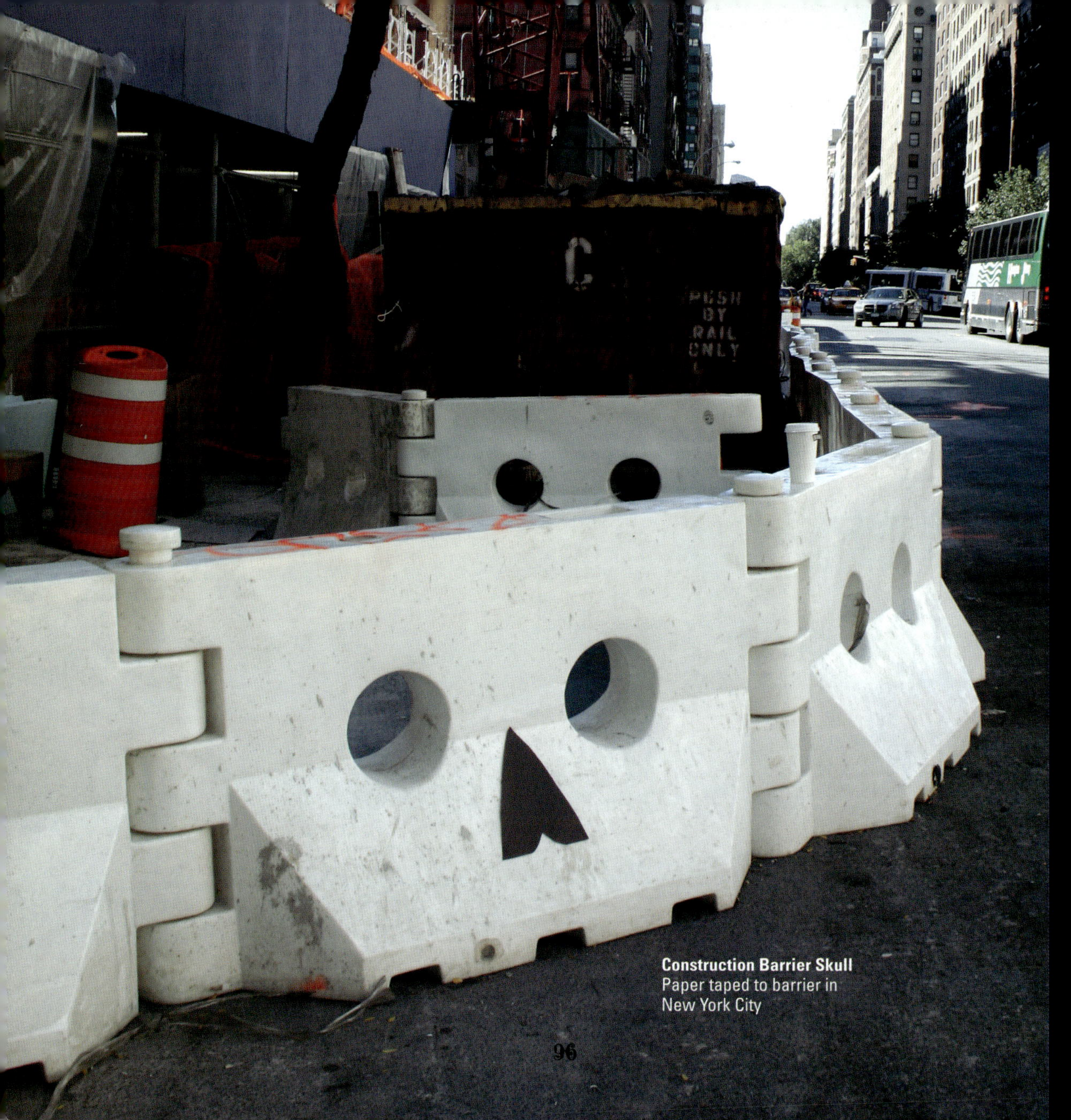

Construction Barrier Skull
Paper taped to barrier in
New York City

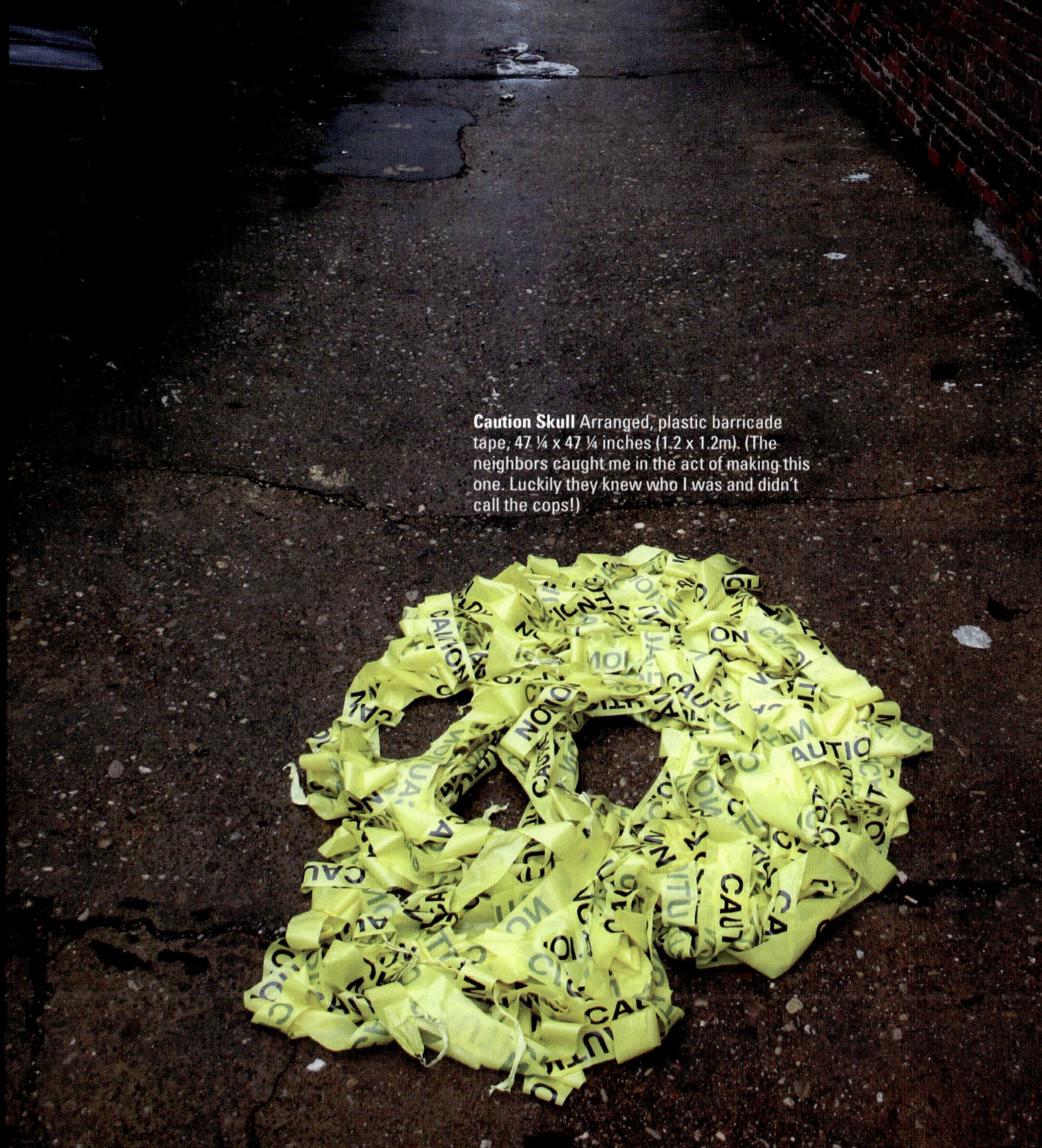

Caution Skull Arranged, plastic barricade tape, 47 ¼ x 47 ¼ inches (1.2 x 1.2m). (The neighbors caught me in the act of making this one. Luckily they knew who I was and didn't call the cops!)

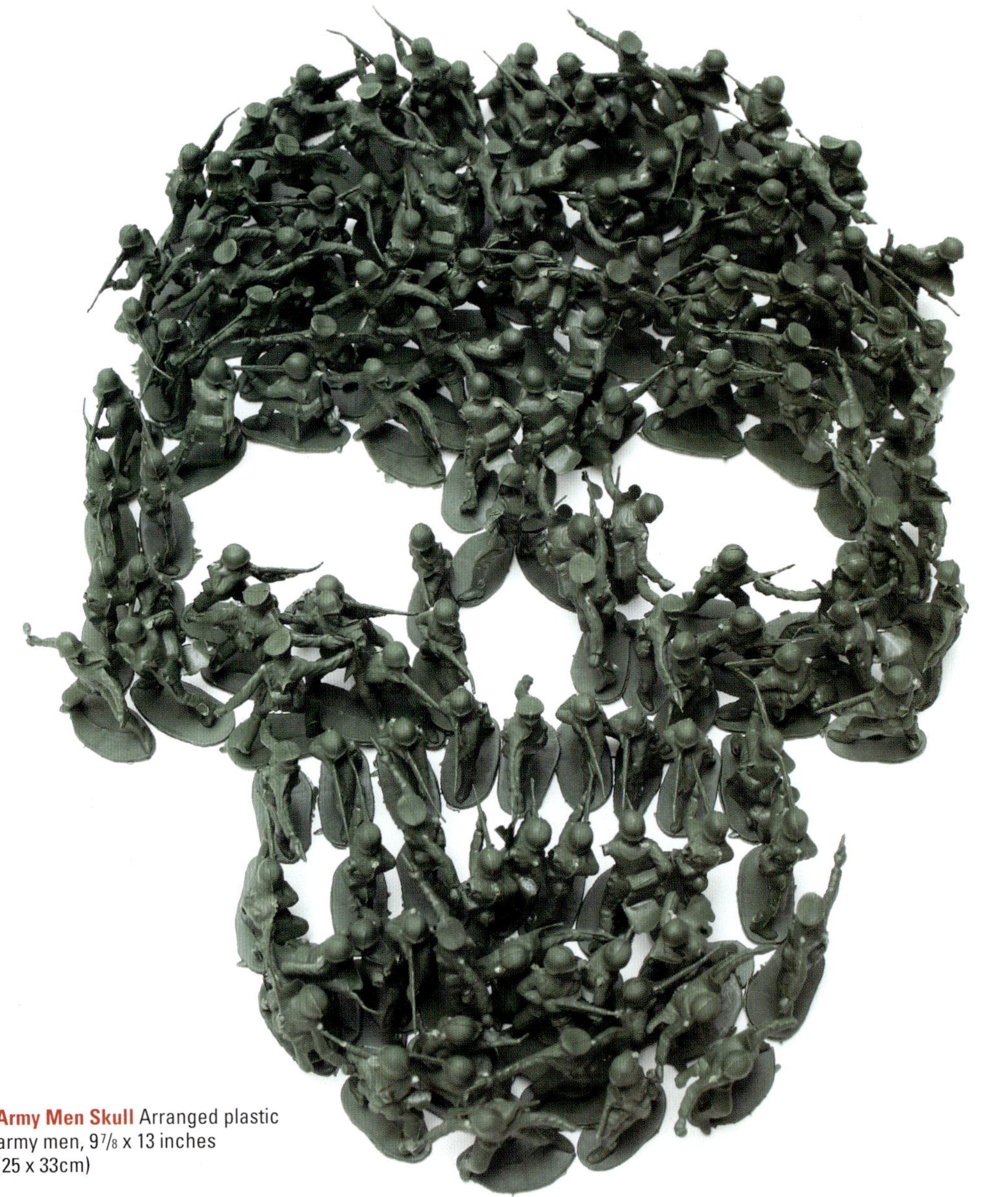

Army Men Skull Arranged plastic army men, $9^{7}/_{8}$ x 13 inches (25 x 33cm)

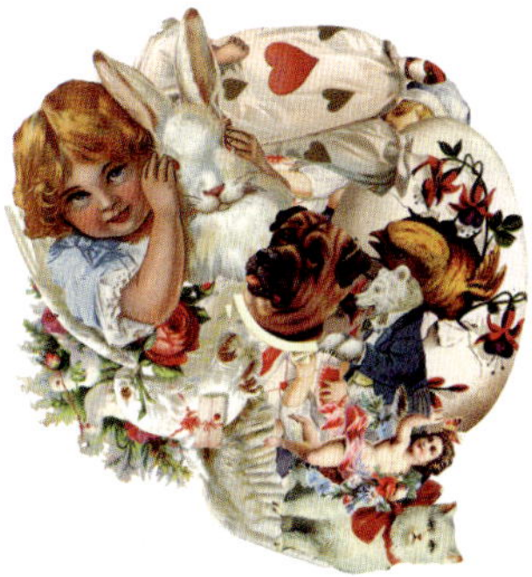

Skullage
Collaged Victorian chromolithographs

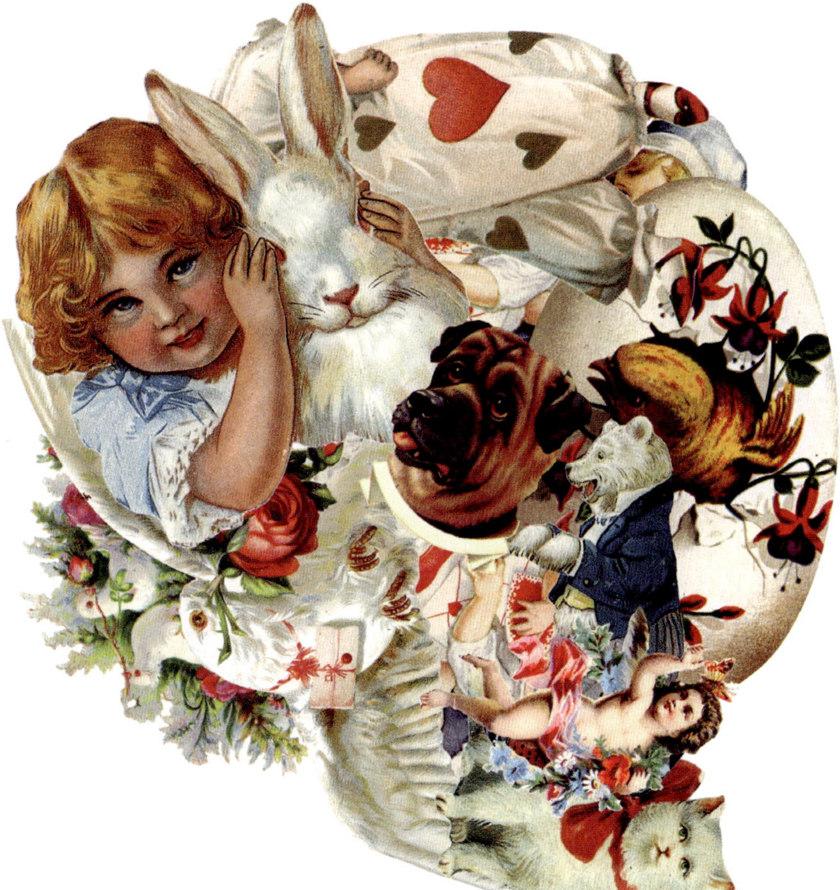

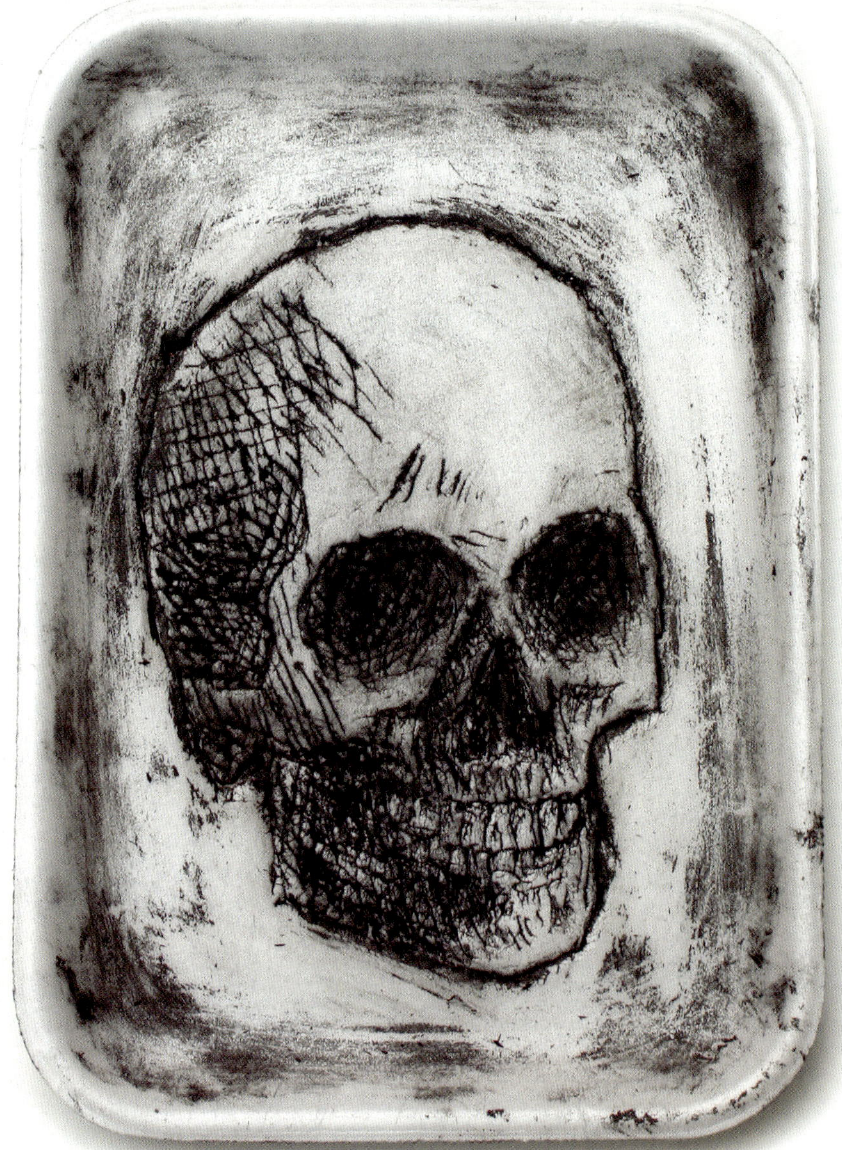

Ink & Tray Skull
Printing ink on etched polystyrene tray, 8¼ x 5¾ inches (21 x 14.6cm). (I hate throwing these things away, and they're difficult to recycle in my neck of the woods, so it's nice to have something to do with them. I try to avoid them when I can, though I'd prefer if they just weren't used anymore!)

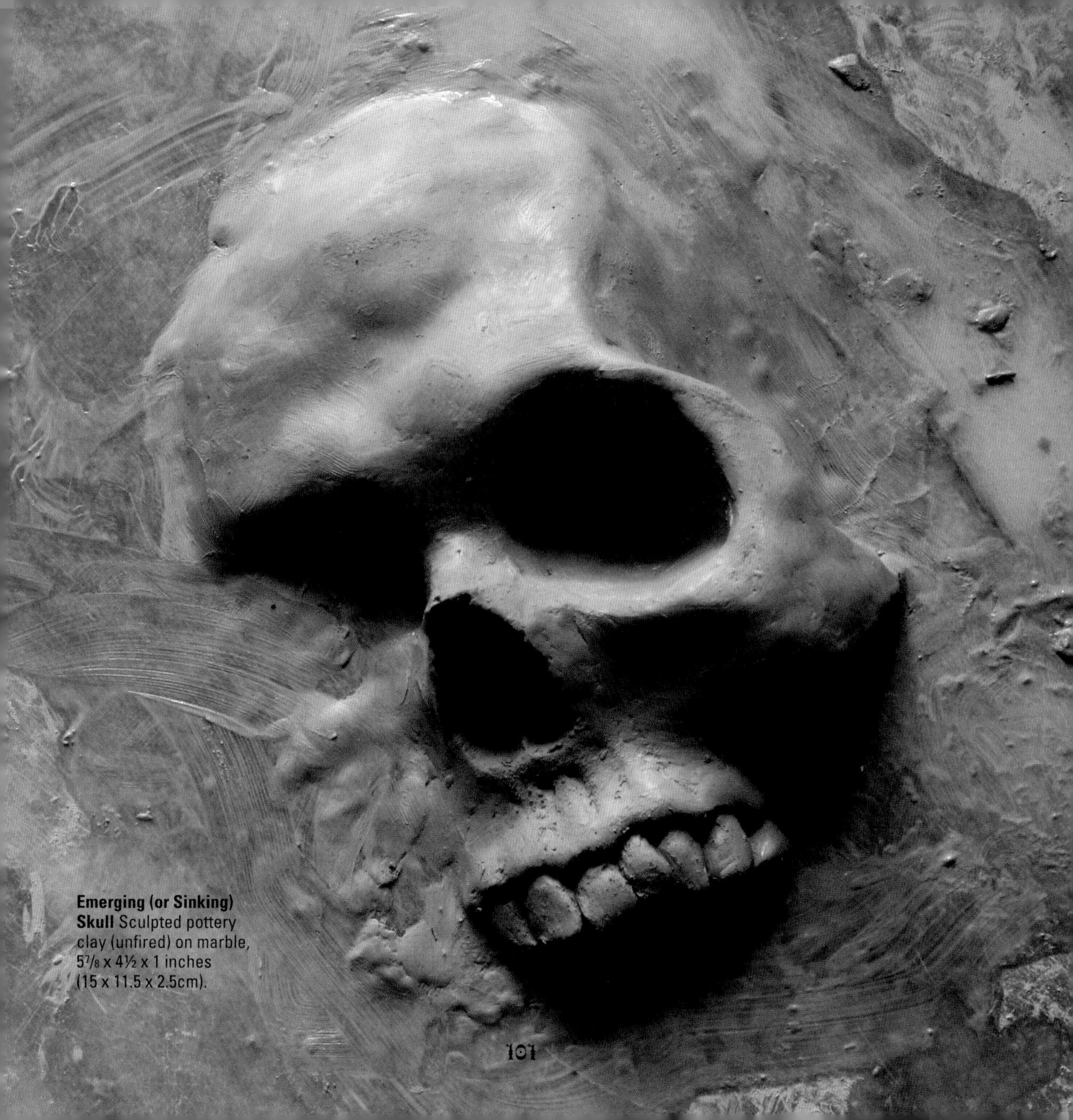

Emerging (or Sinking) Skull Sculpted pottery clay (unfired) on marble, 5⅞ x 4½ x 1 inches (15 x 11.5 x 2.5cm).

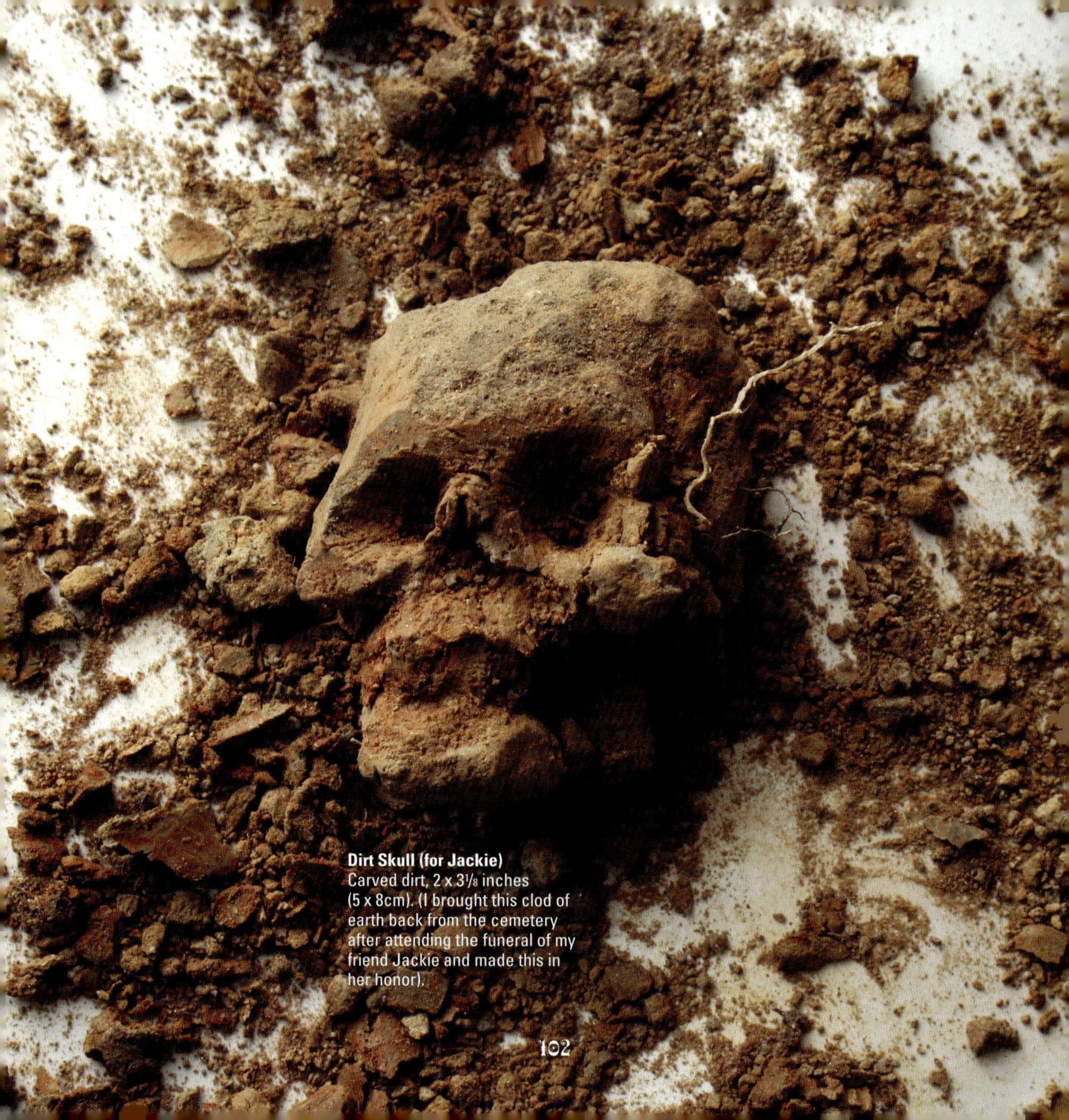

Dirt Skull (for Jackie)
Carved dirt, 2 x 3⅛ inches (5 x 8cm). (I brought this clod of earth back from the cemetery after attending the funeral of my friend Jackie and made this in her honor).

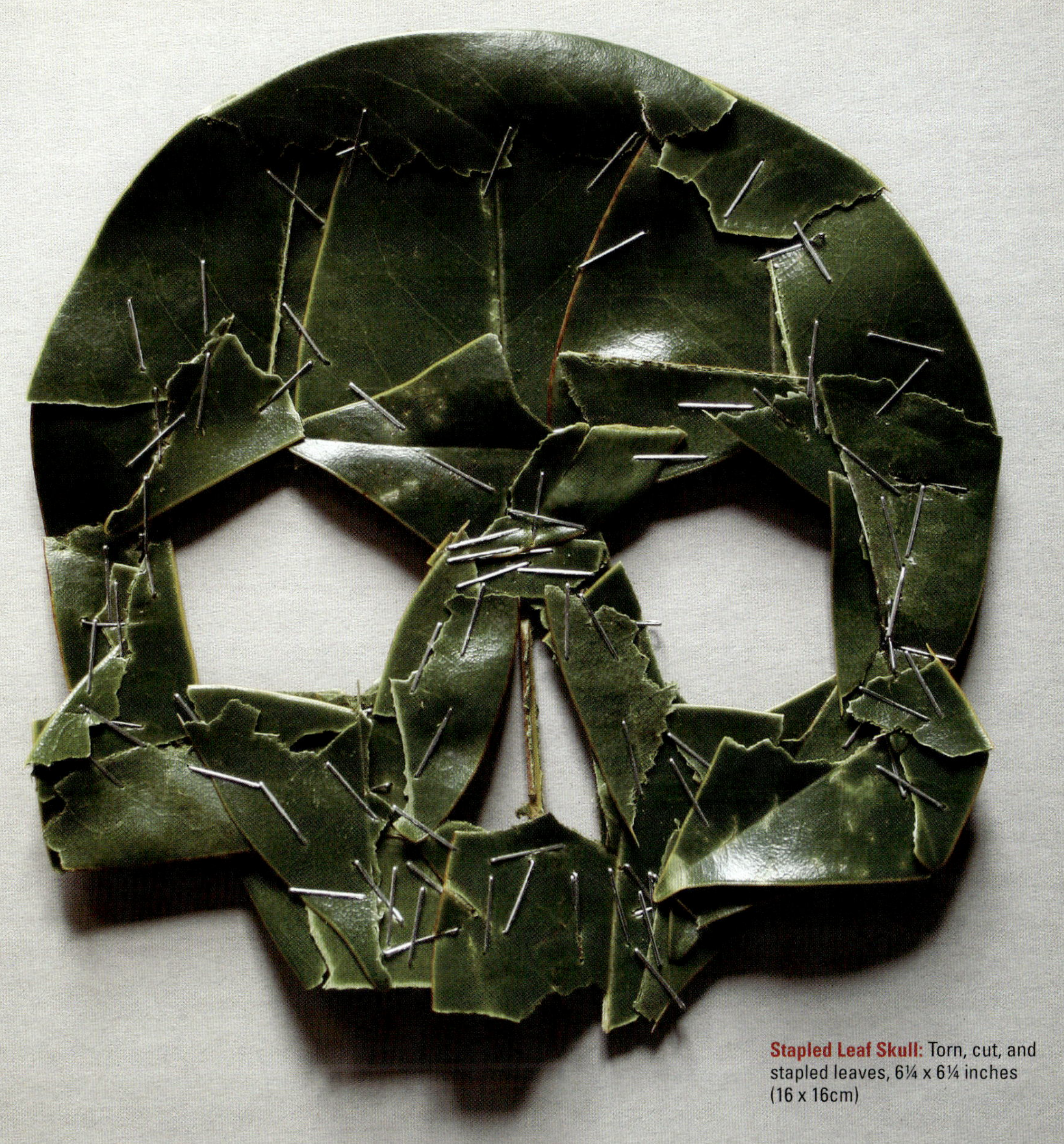

Stapled Leaf Skull: Torn, cut, and stapled leaves, 6¼ x 6¼ inches (16 x 16cm)

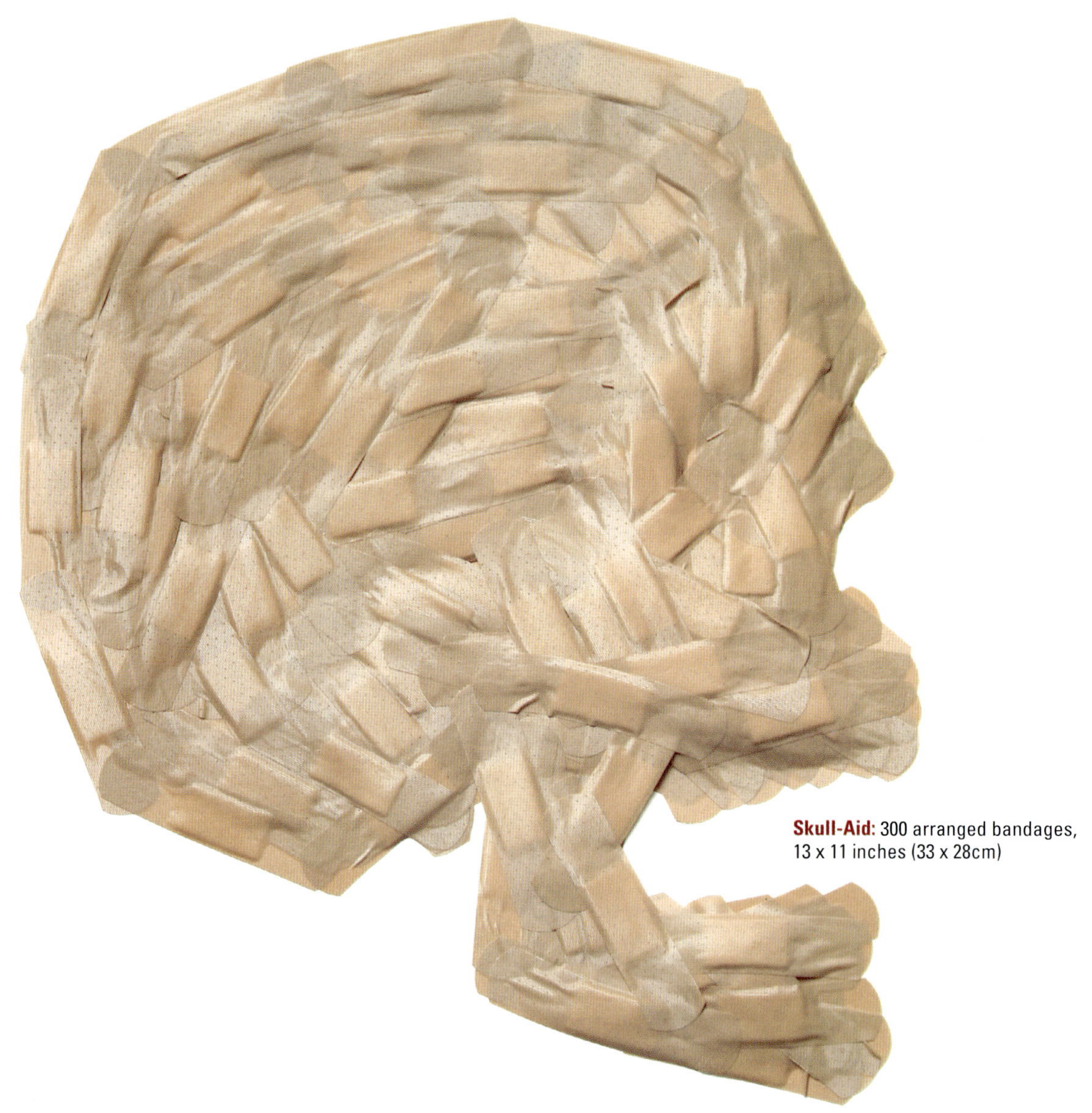

Skull-Aid: 300 arranged bandages, 13 x 11 inches (33 x 28cm)

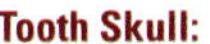

Tooth Skull:
Carved human tooth.
(Yes, this is one of my actual wisdom teeth that I've kept since it was pulled when I was in high school.)

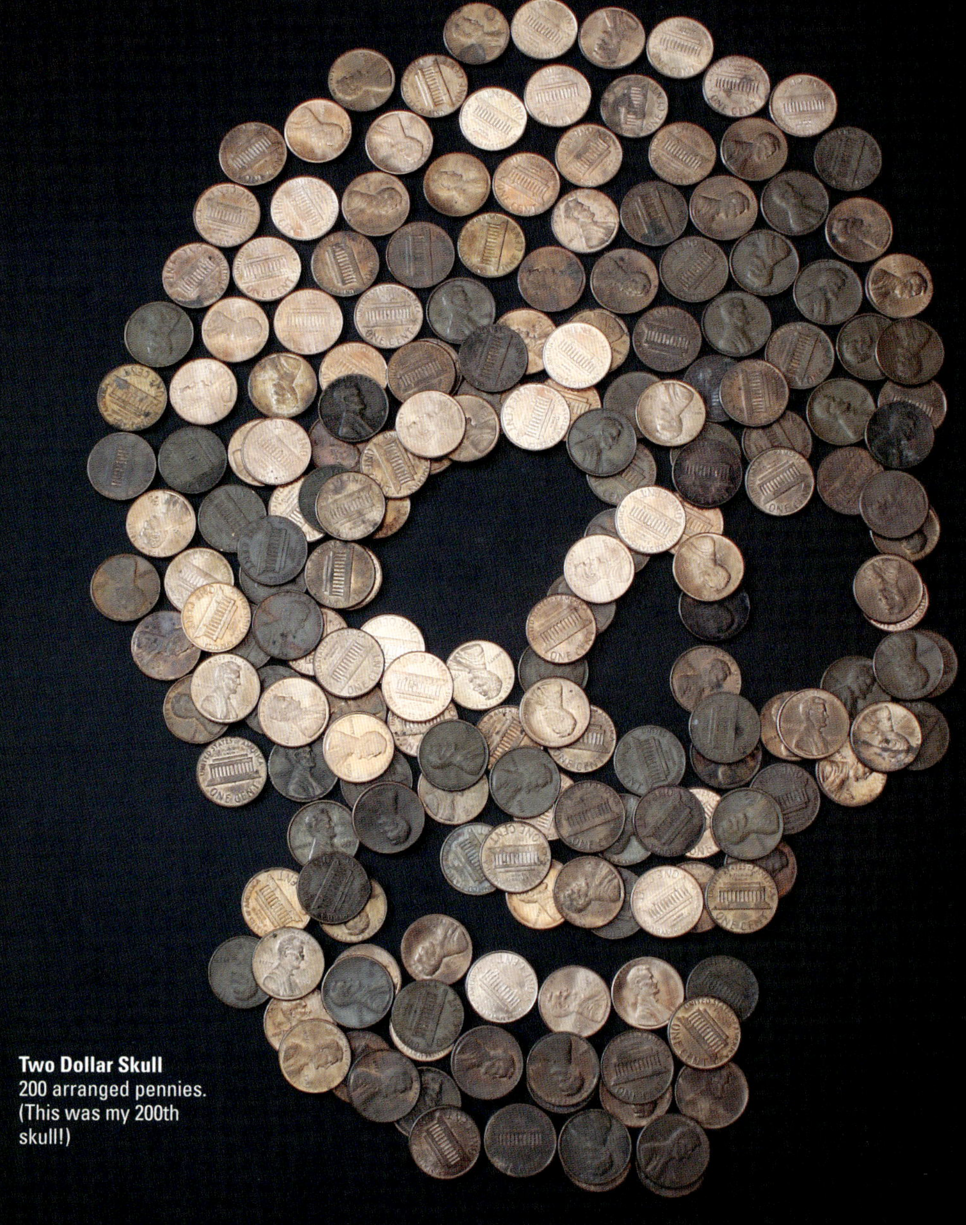

Two Dollar Skull
200 arranged pennies.
(This was my 200th
skull!)

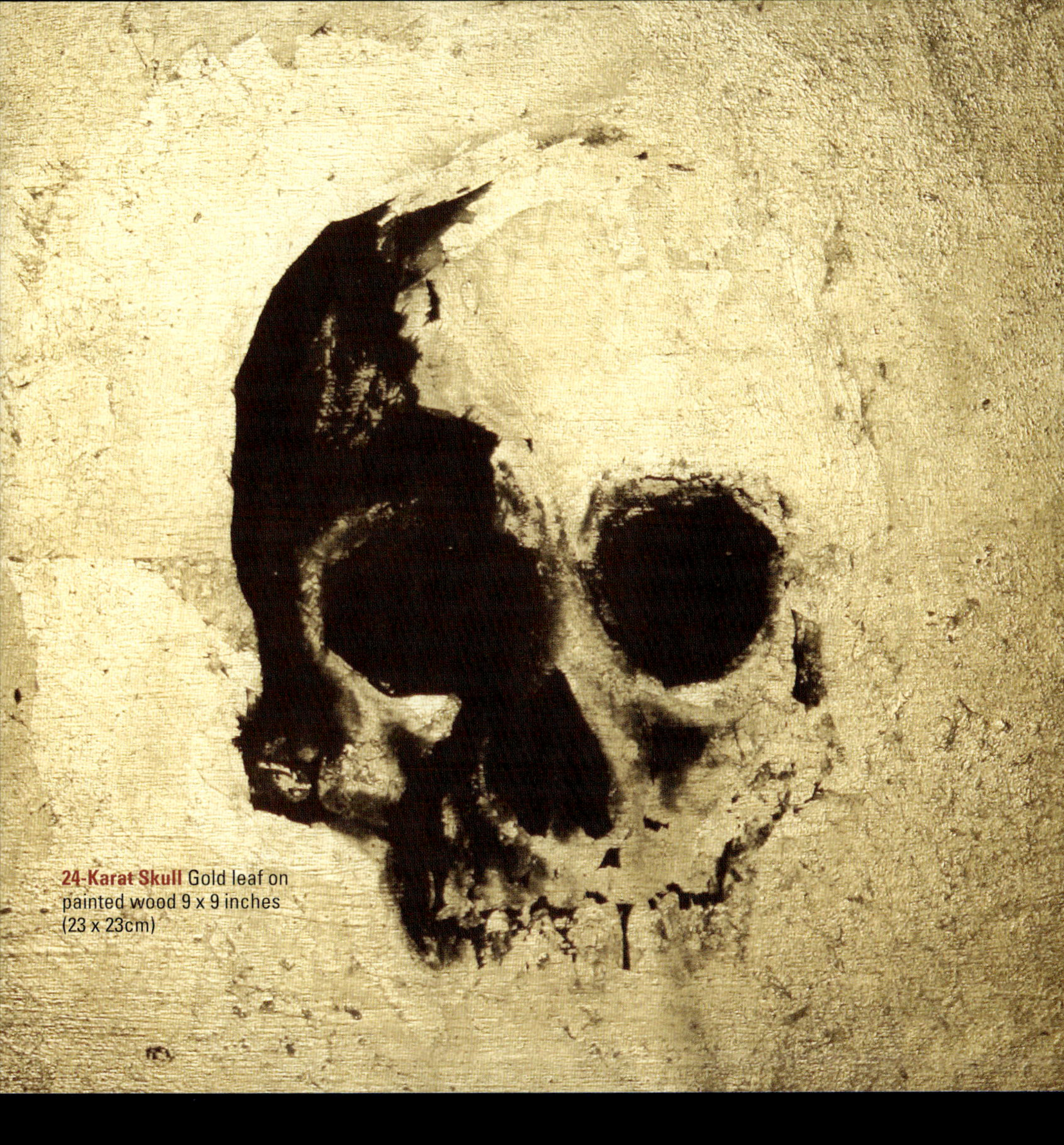

24-Karat Skull Gold leaf on painted wood 9 x 9 inches (23 x 23cm)

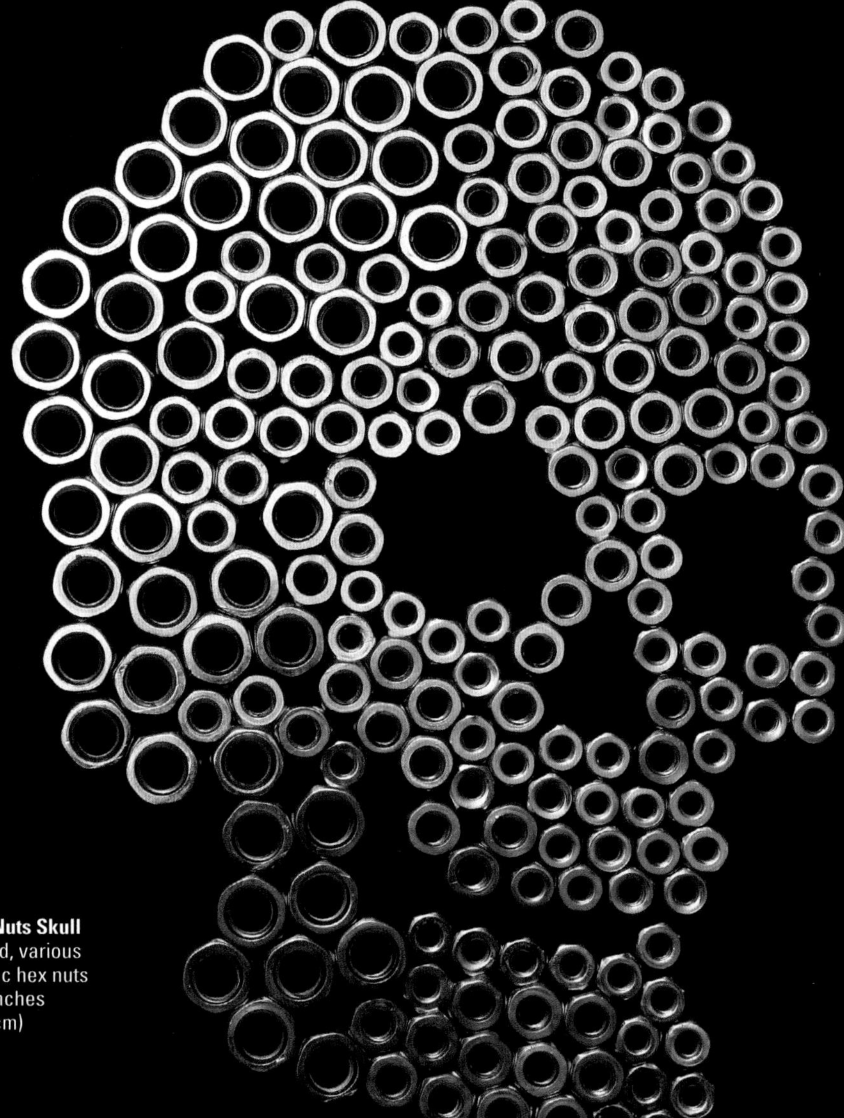

Totally Nuts Skull
Arranged, various sized zinc hex nuts
7 1/8 x 9 inches
(18 x 23cm)

Rorshach Skull Digital ink-blot illustration

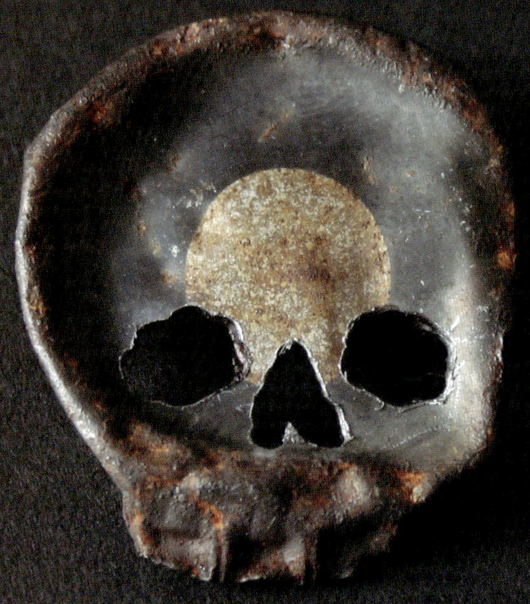

Bottle Cap Skull Cut metal bottle cap. (I found this on the ground already bent into this shape.)

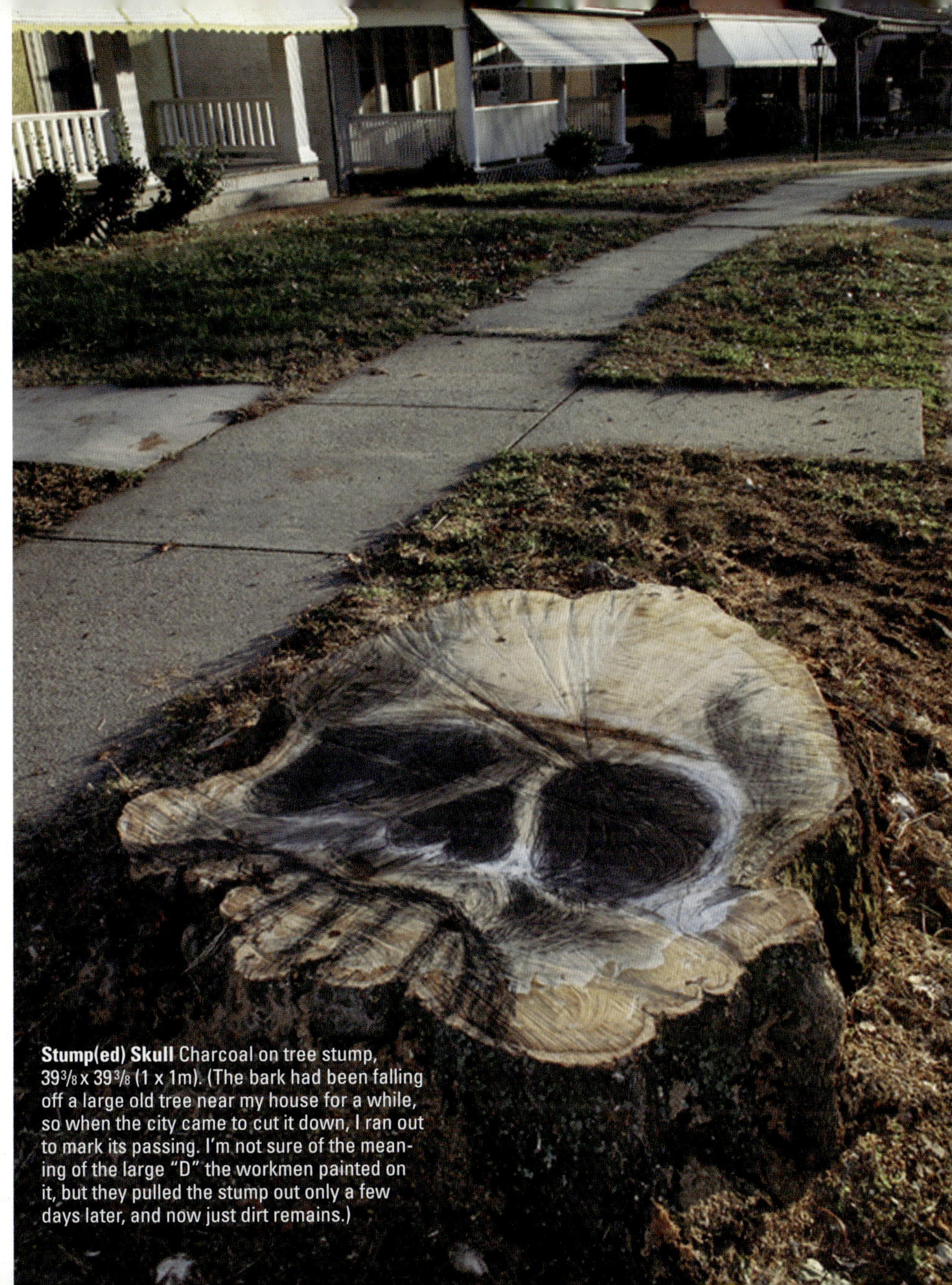

Stump(ed) Skull Charcoal on tree stump, $39^3/_8$ x $39^3/_8$ (1 x 1m). (The bark had been falling off a large old tree near my house for a while, so when the city came to cut it down, I ran out to mark its passing. I'm not sure of the meaning of the large "D" the workmen painted on it, but they pulled the stump out only a few days later, and now just dirt remains.)

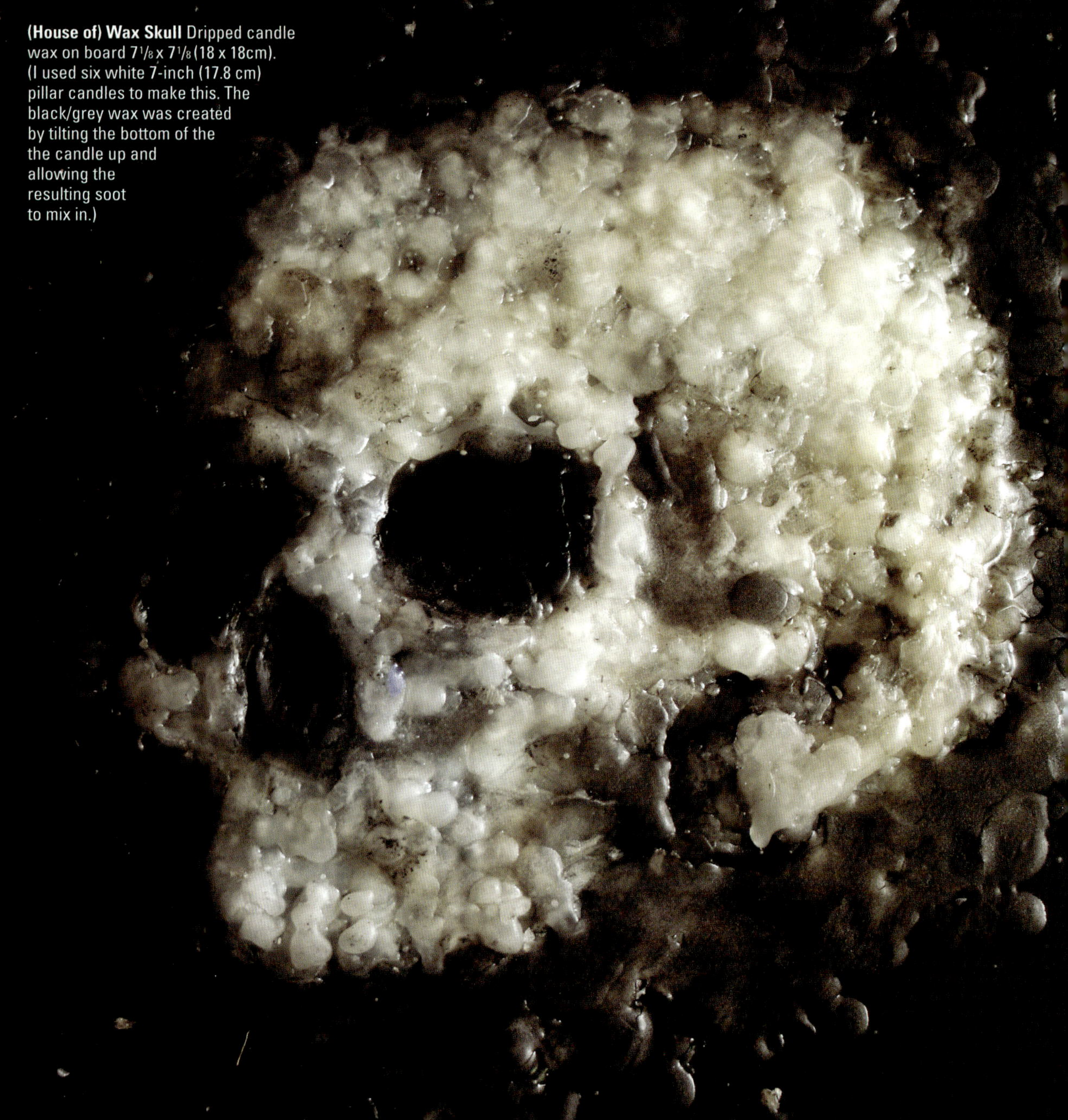

(House of) Wax Skull Dripped candle wax on board 7 1/8 x 7 1/8 (18 x 18cm). (I used six white 7-inch (17.8 cm) pillar candles to make this. The black/grey wax was created by tilting the bottom of the the candle up and allowing the resulting soot to mix in.)

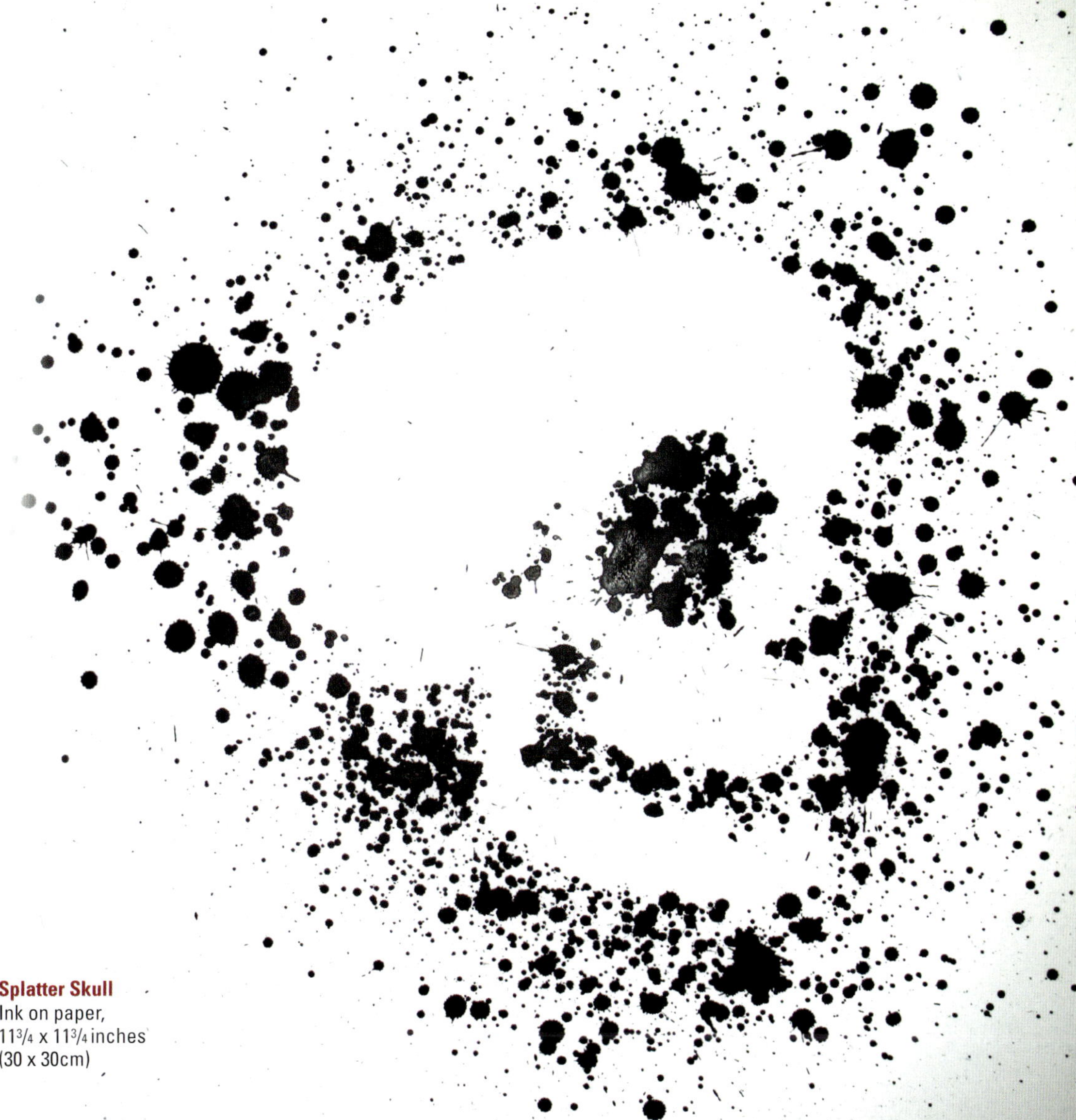

Splatter Skull
Ink on paper,
$11^3/_4$ x $11^3/_4$ inches
(30 x 30cm)

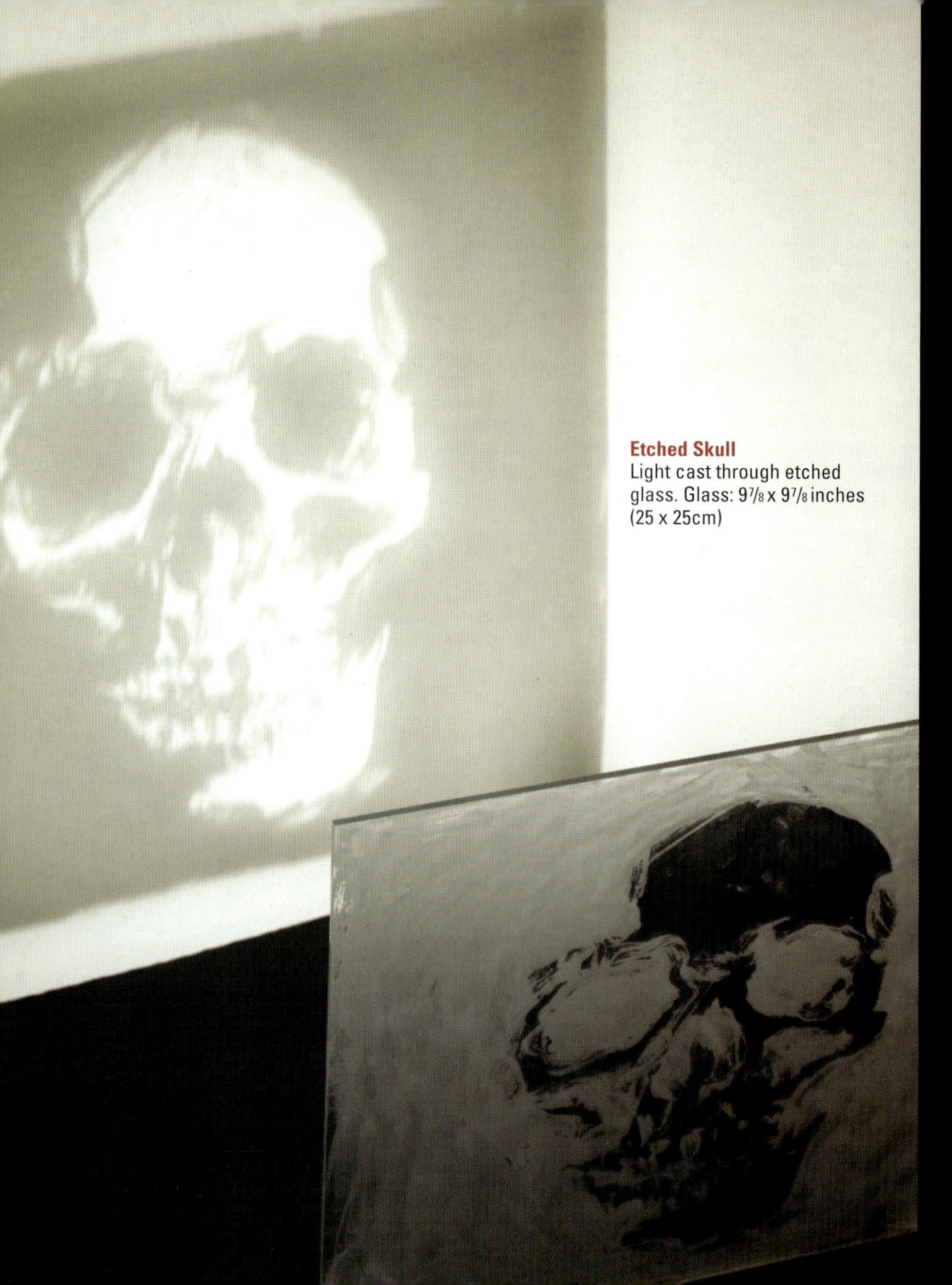

Etched Skull
Light cast through etched glass. Glass: 9⅞ x 9⅞ inches (25 x 25cm)

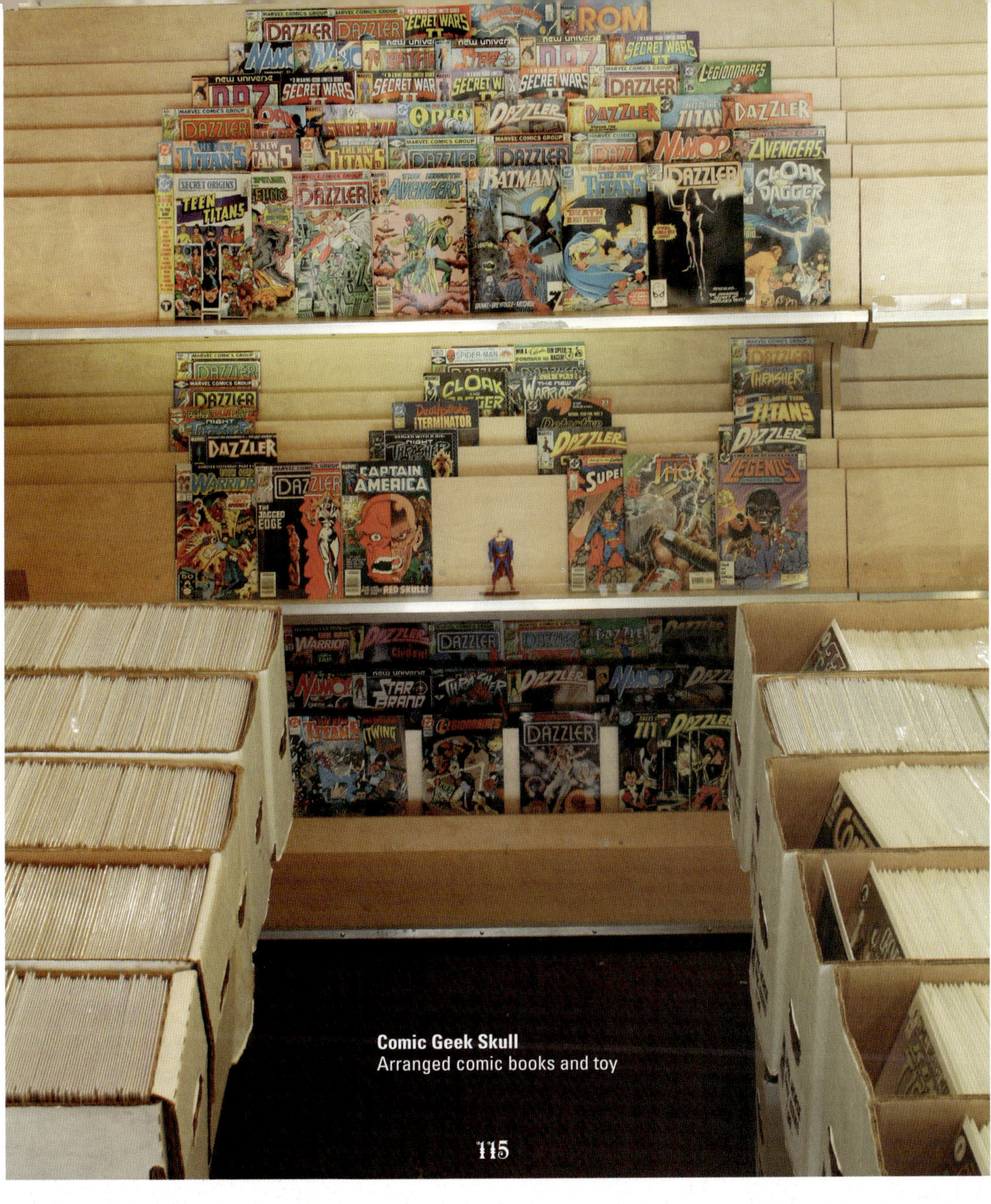

Comic Geek Skull
Arranged comic books and toy

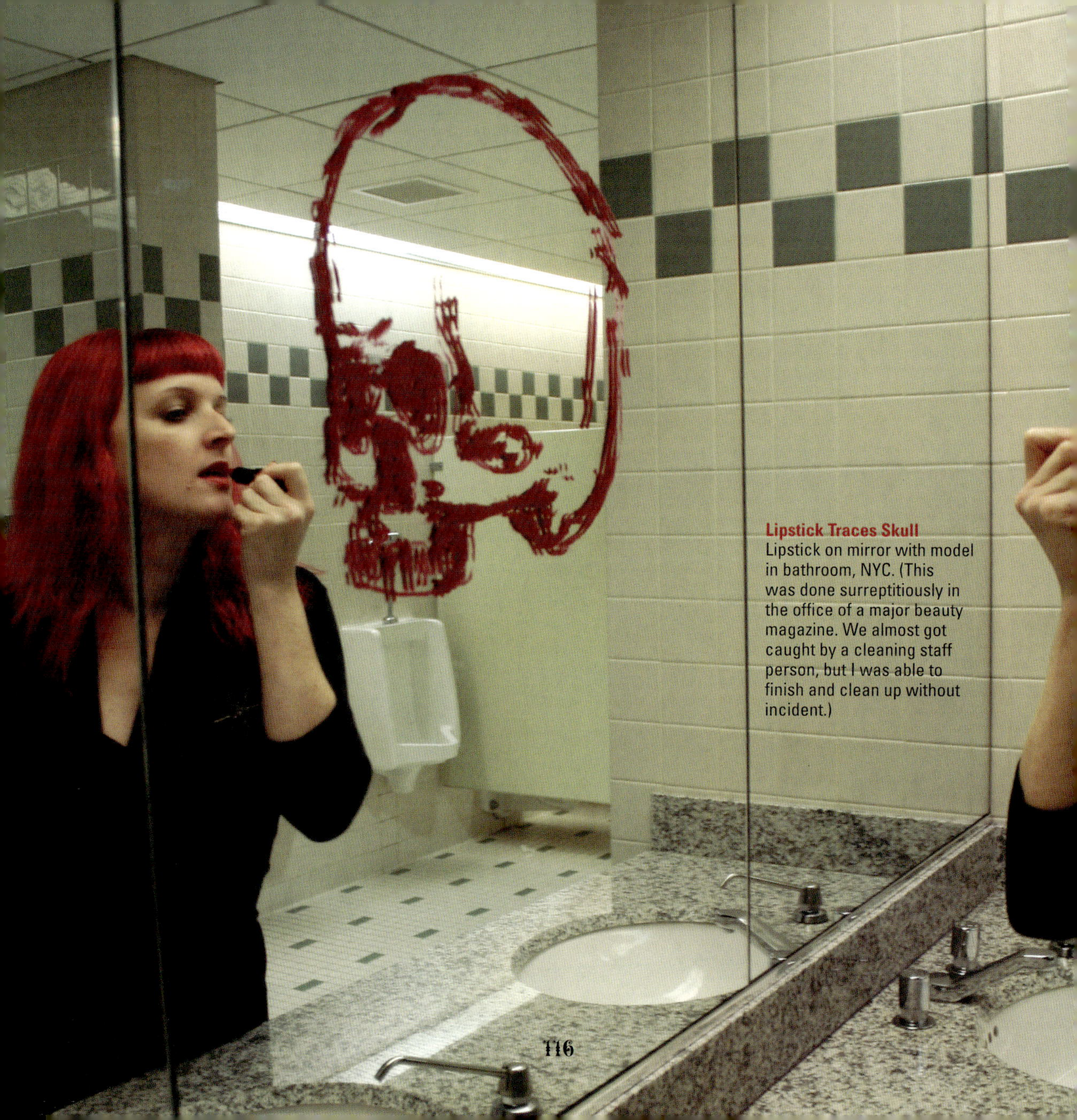

Lipstick Traces Skull
Lipstick on mirror with model in bathroom, NYC. (This was done surreptitiously in the office of a major beauty magazine. We almost got caught by a cleaning staff person, but I was able to finish and clean up without incident.)

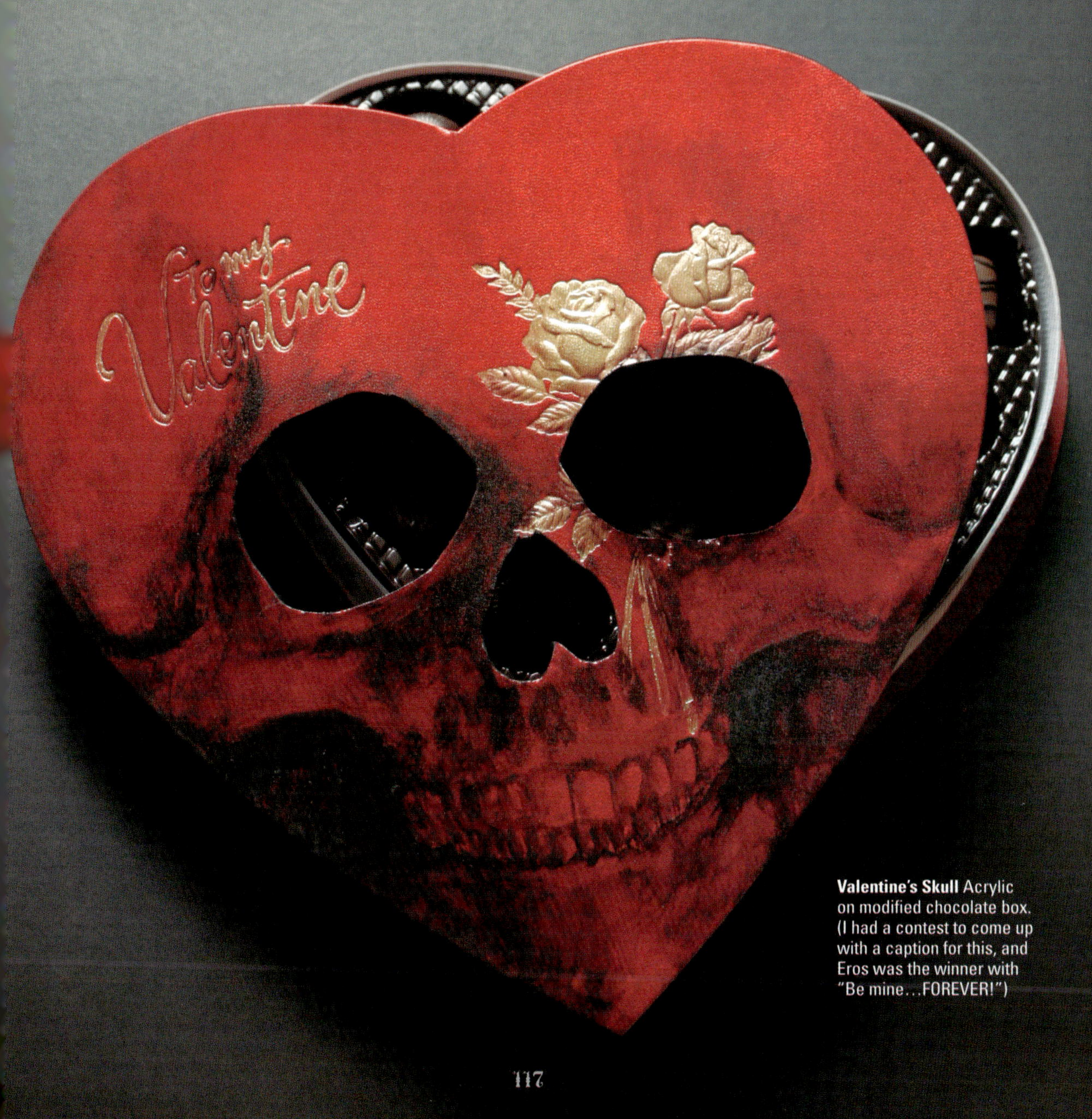

Valentine's Skull Acrylic on modified chocolate box. (I had a contest to come up with a caption for this, and Eros was the winner with "Be mine...FOREVER!")

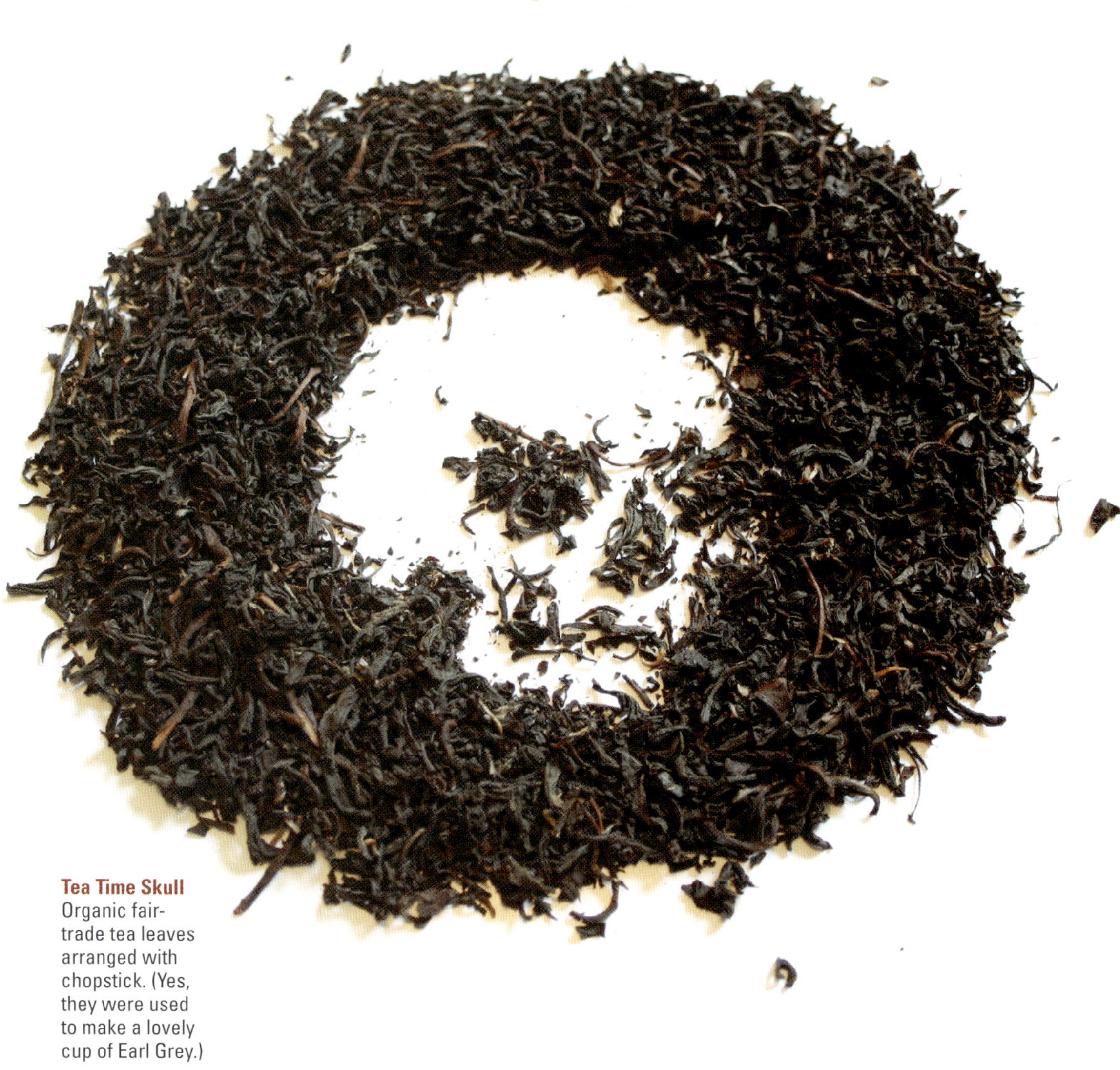

Tea Time Skull
Organic fair-trade tea leaves arranged with chopstick. (Yes, they were used to make a lovely cup of Earl Grey.)

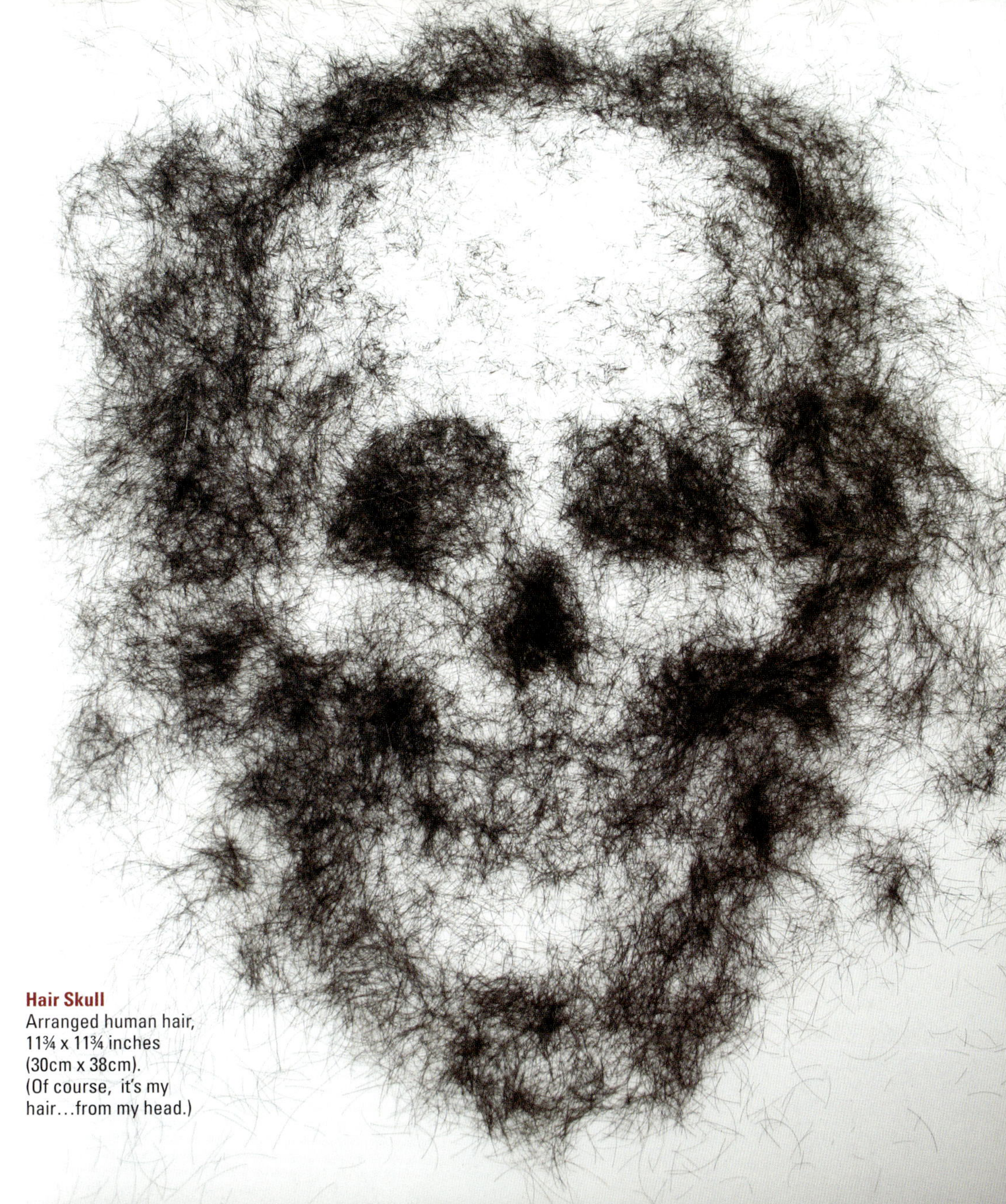

Hair Skull
Arranged human hair,
11¾ x 11¾ inches
(30cm x 38cm).
(Of course, it's my
hair…from my head.)

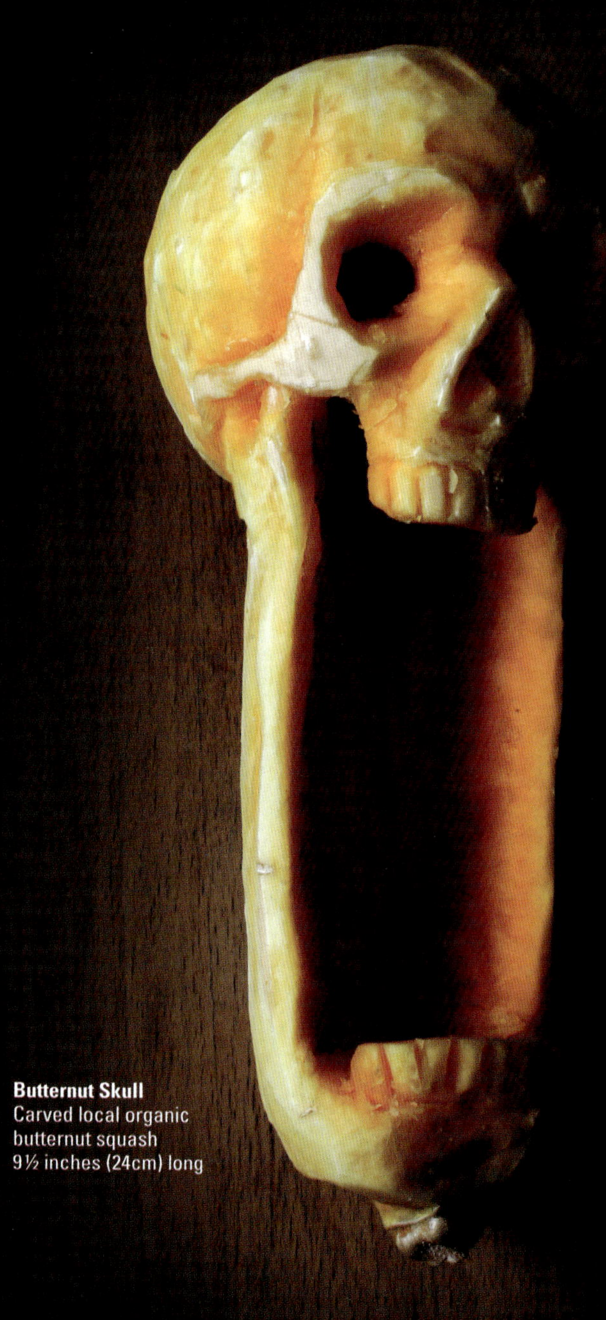

Butternut Skull
Carved local organic
butternut squash
9½ inches (24cm) long

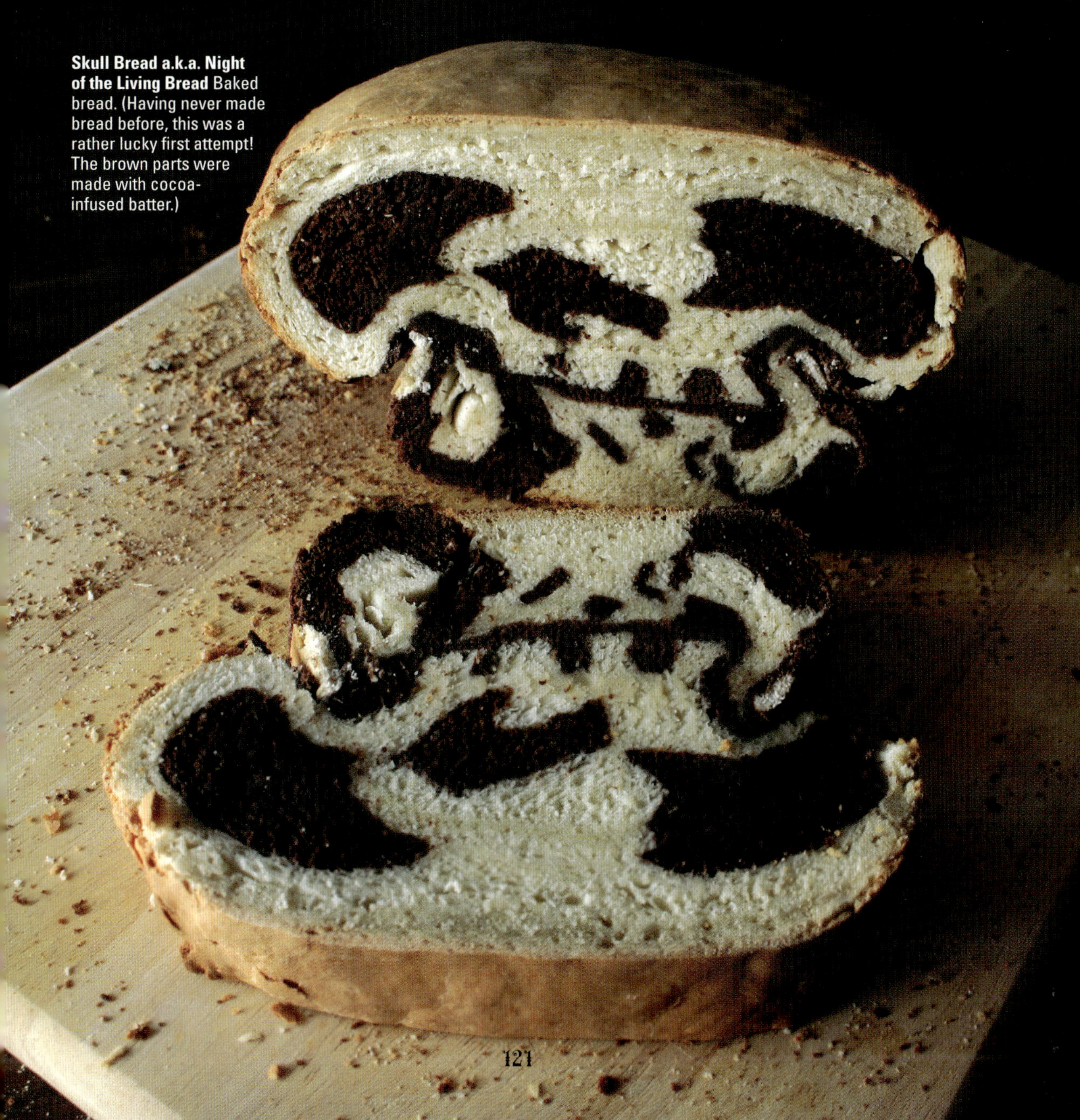

Skull Bread a.k.a. Night of the Living Bread Baked bread. (Having never made bread before, this was a rather lucky first attempt! The brown parts were made with cocoa-infused batter.)

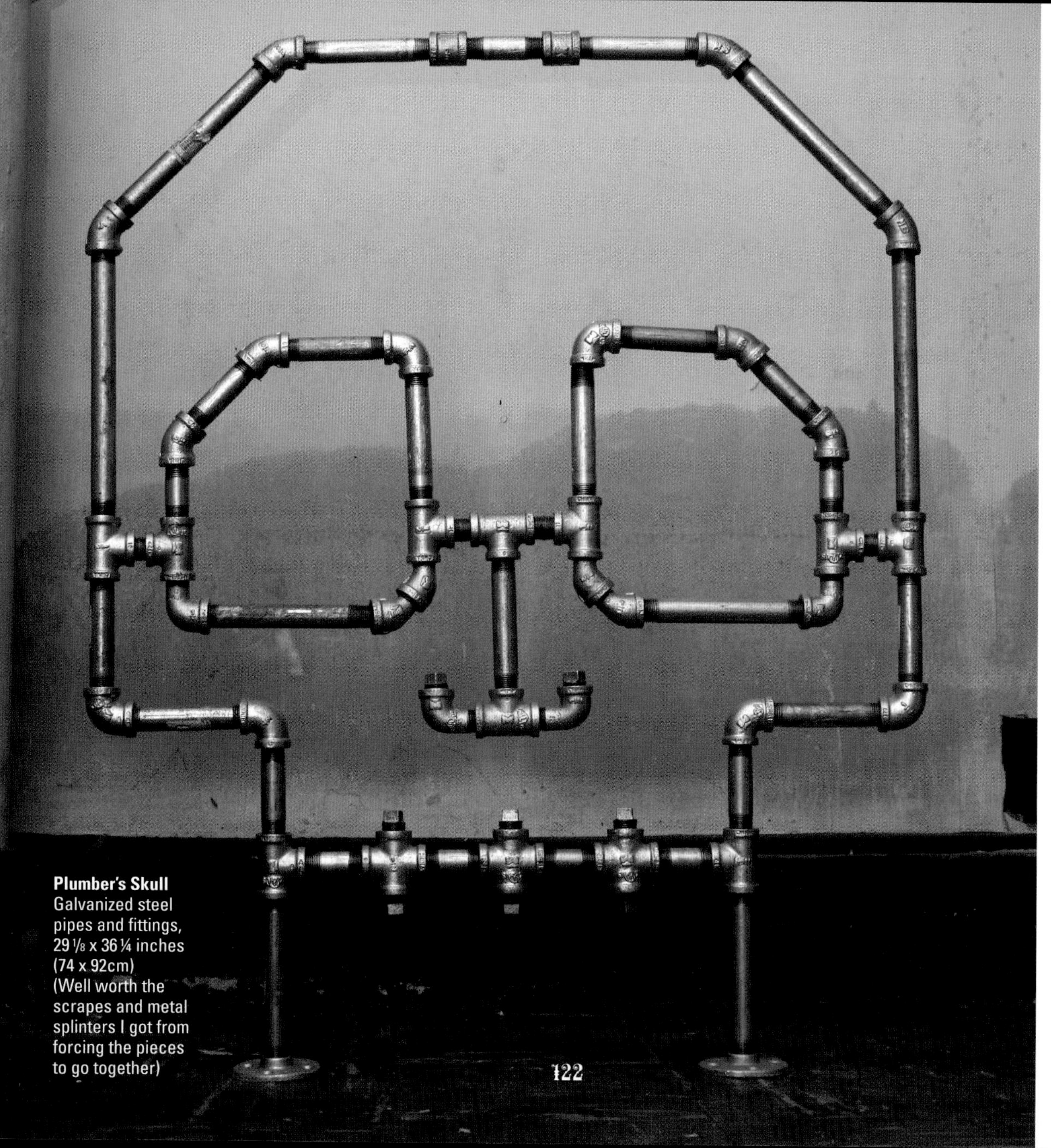

Plumber's Skull
Galvanized steel pipes and fittings, 29 1/8 x 36 1/4 inches (74 x 92cm)
(Well worth the scrapes and metal splinters I got from forcing the pieces to go together)

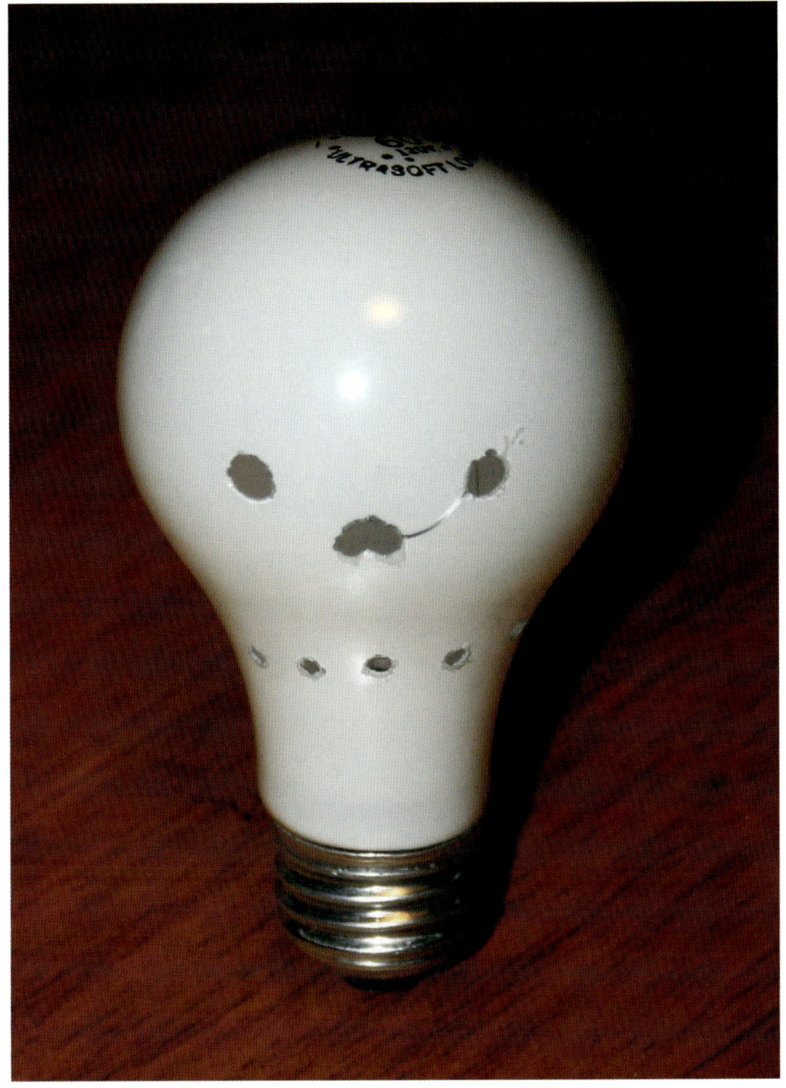

Skull Bulb
Holes drilled in light bulb. (Not an easy task. The only survivor of eight attempts.)

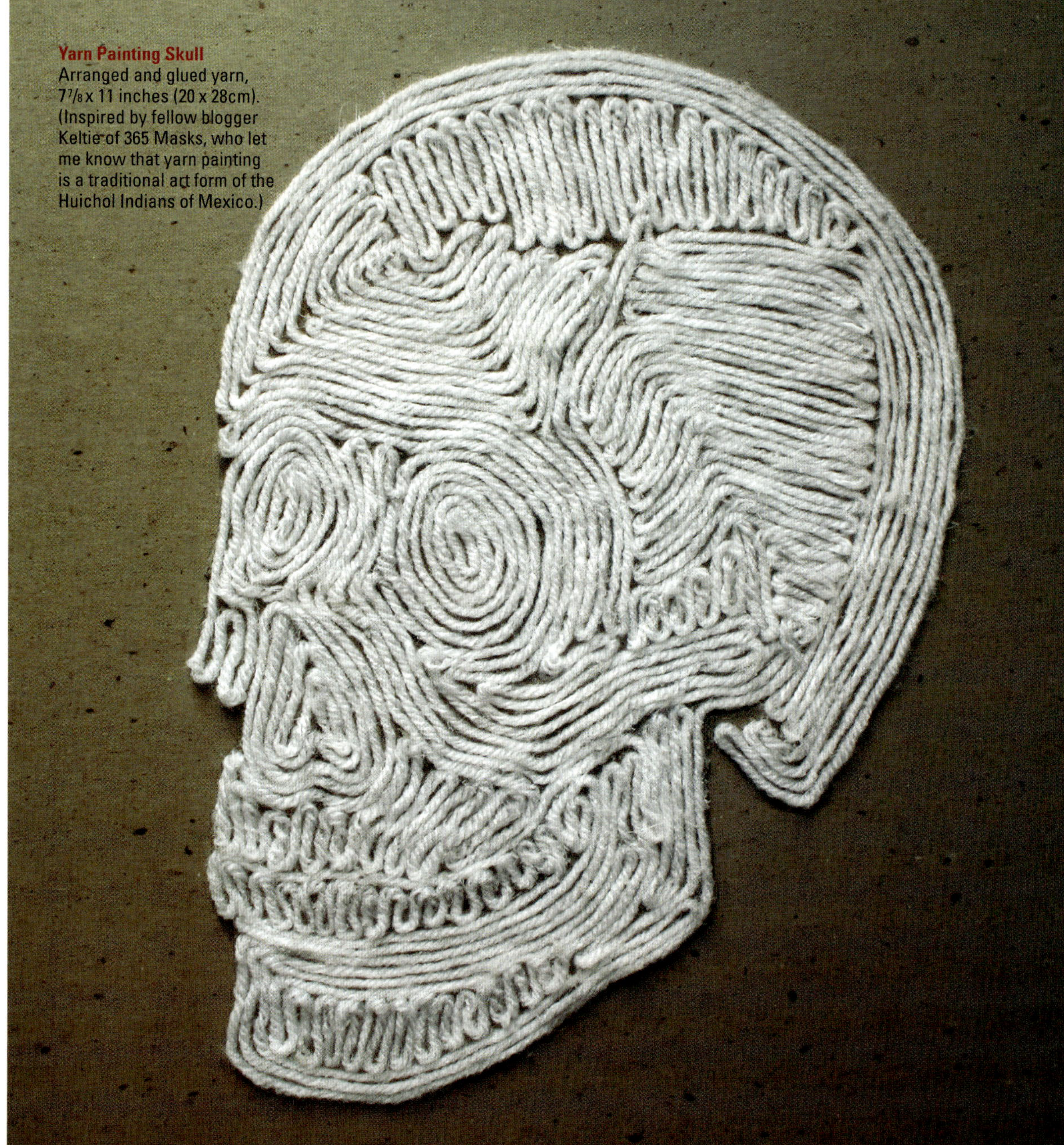

Yarn Painting Skull
Arranged and glued yarn, $7^7/_8$ x 11 inches (20 x 28cm). (Inspired by fellow blogger Keltie of 365 Masks, who let me know that yarn painting is a traditional art form of the Huichol Indians of Mexico.)

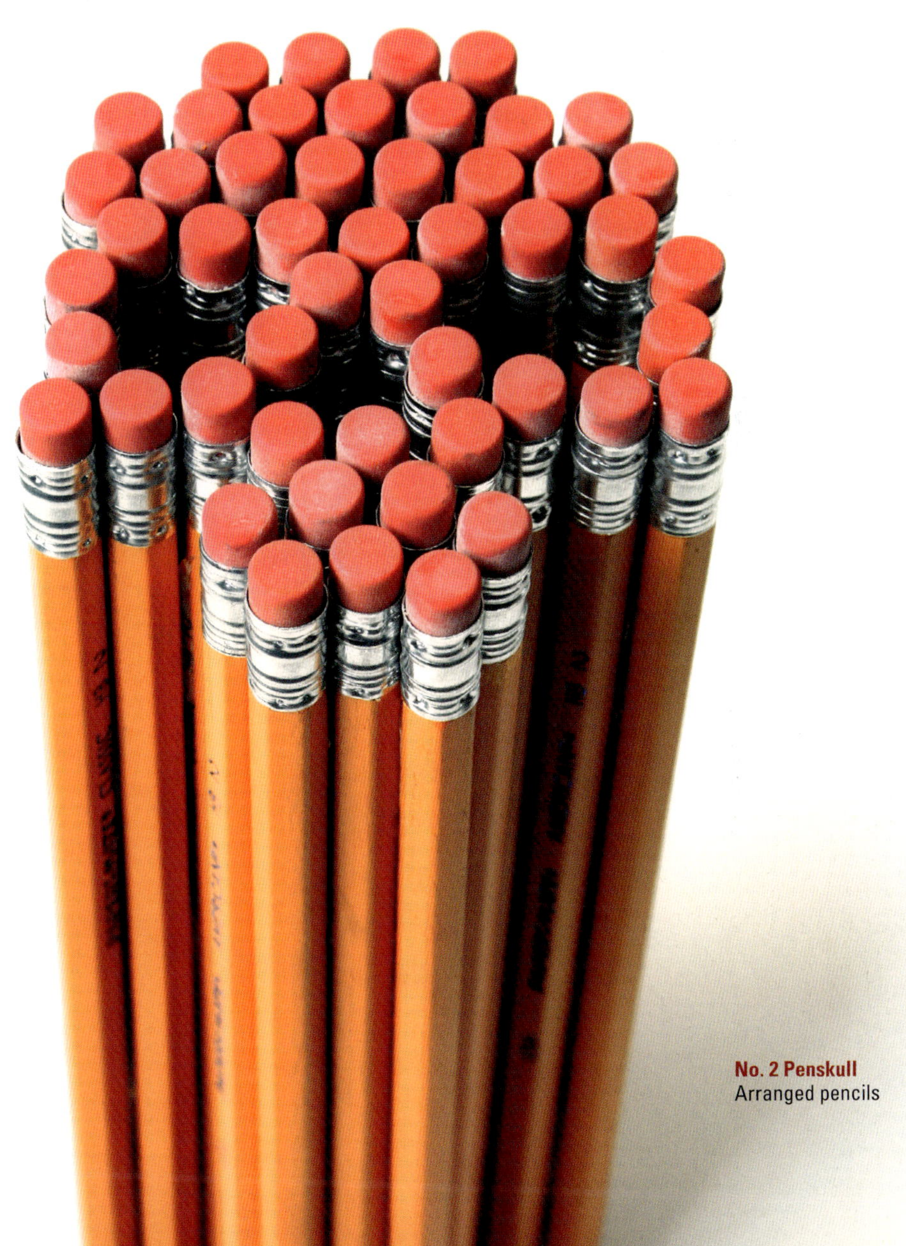

No. 2 Penskull
Arranged pencils

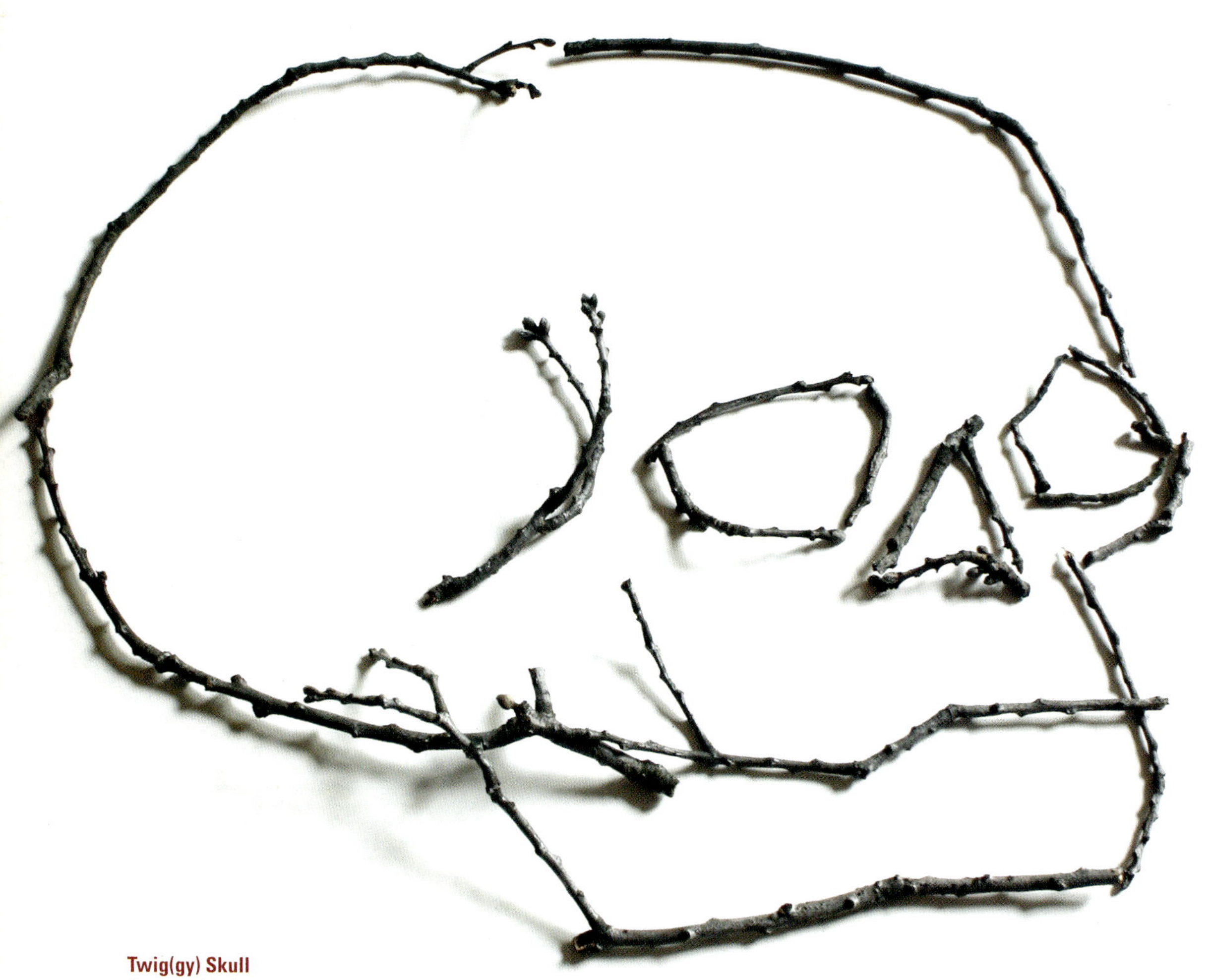

Twig(gy) Skull
Arranged twigs,
15 x 15 inches (38 x 38cm).
(A storm left these twigs
for me in front of my house.)

Five Golden Skulls One length of hand-bent 24-gauge metal wire, 4½ x 7⅛ inches (11.5 x 18cm)

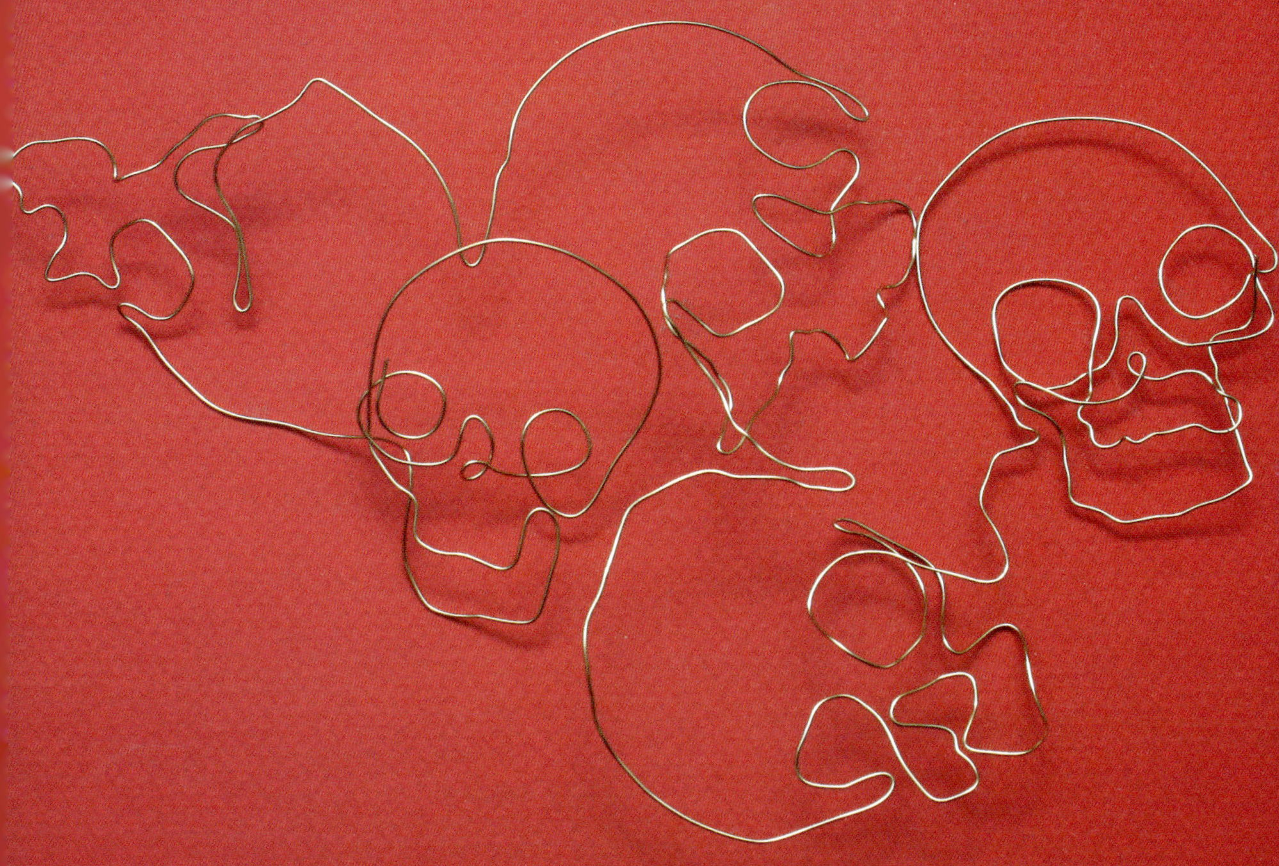

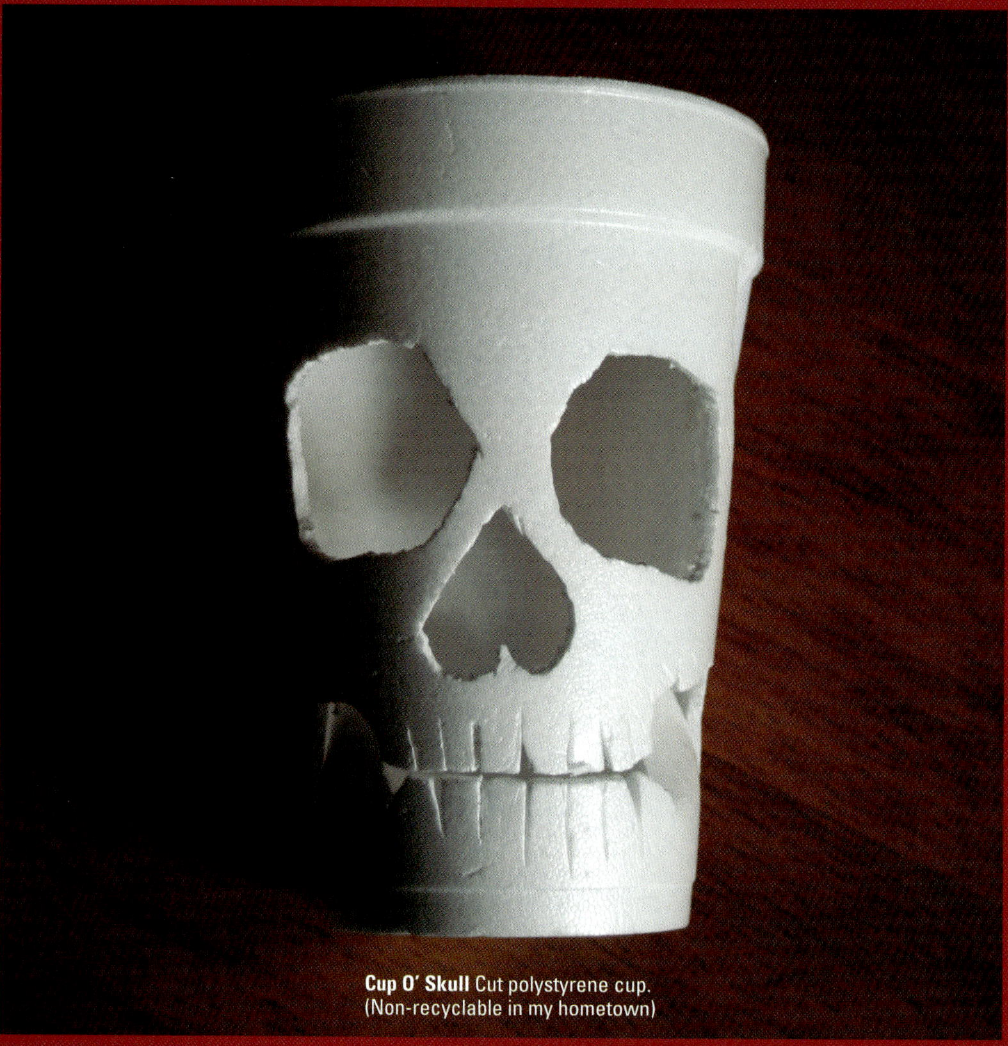

Cup O' Skull Cut polystyrene cup.
(Non-recyclable in my hometown)

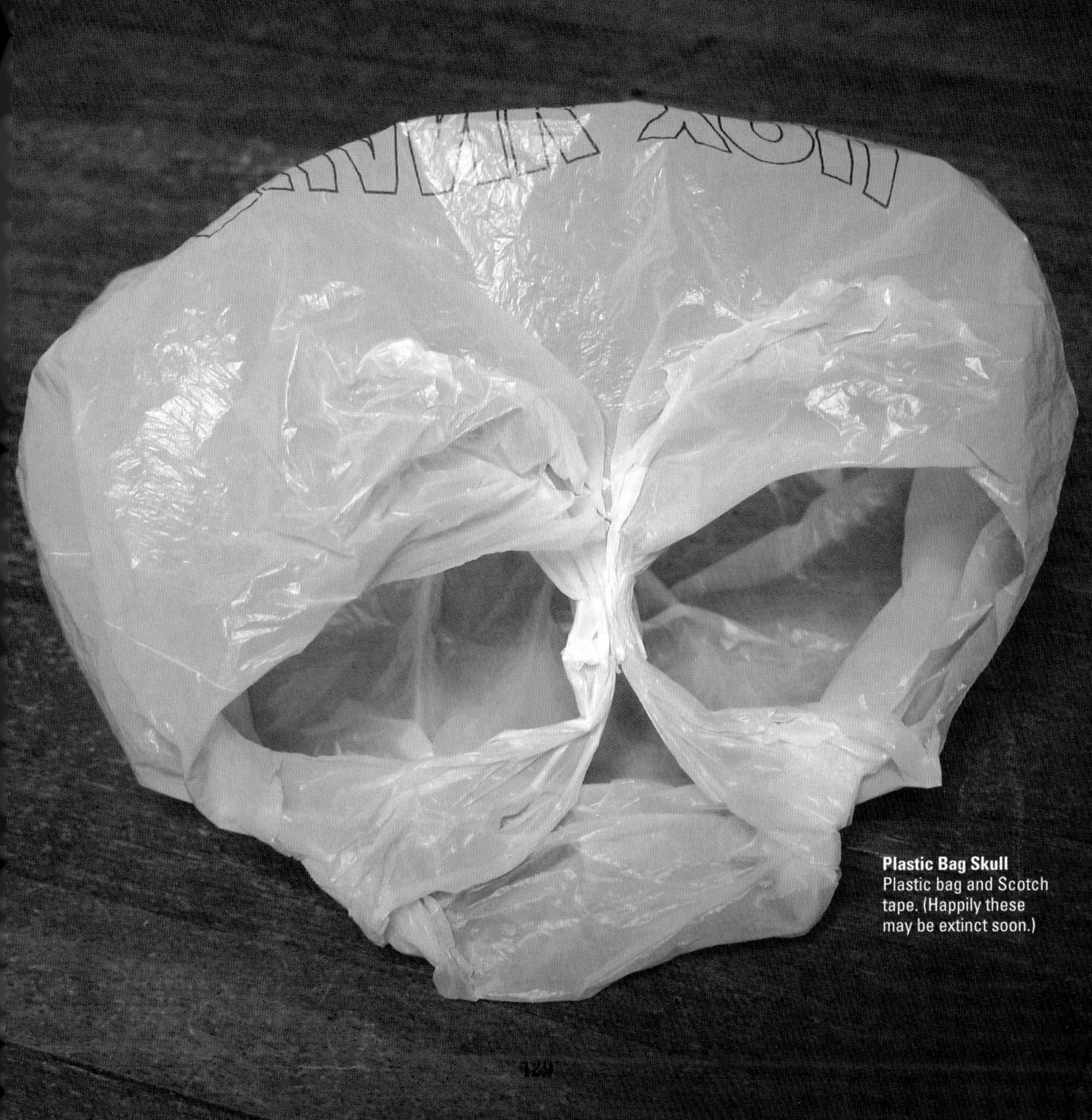

Plastic Bag Skull
Plastic bag and Scotch tape. (Happily these may be extinct soon.)

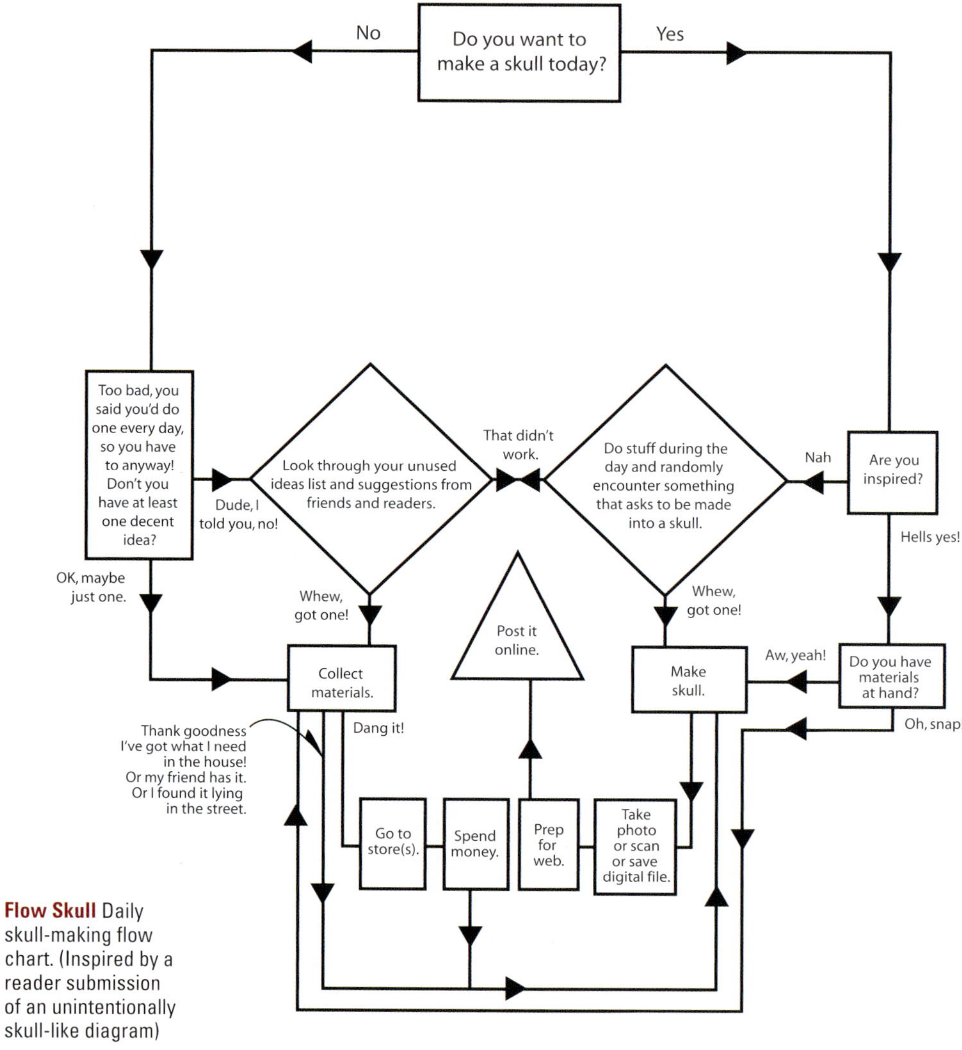

Flow Skull Daily skull-making flow chart. (Inspired by a reader submission of an unintentionally skull-like diagram)

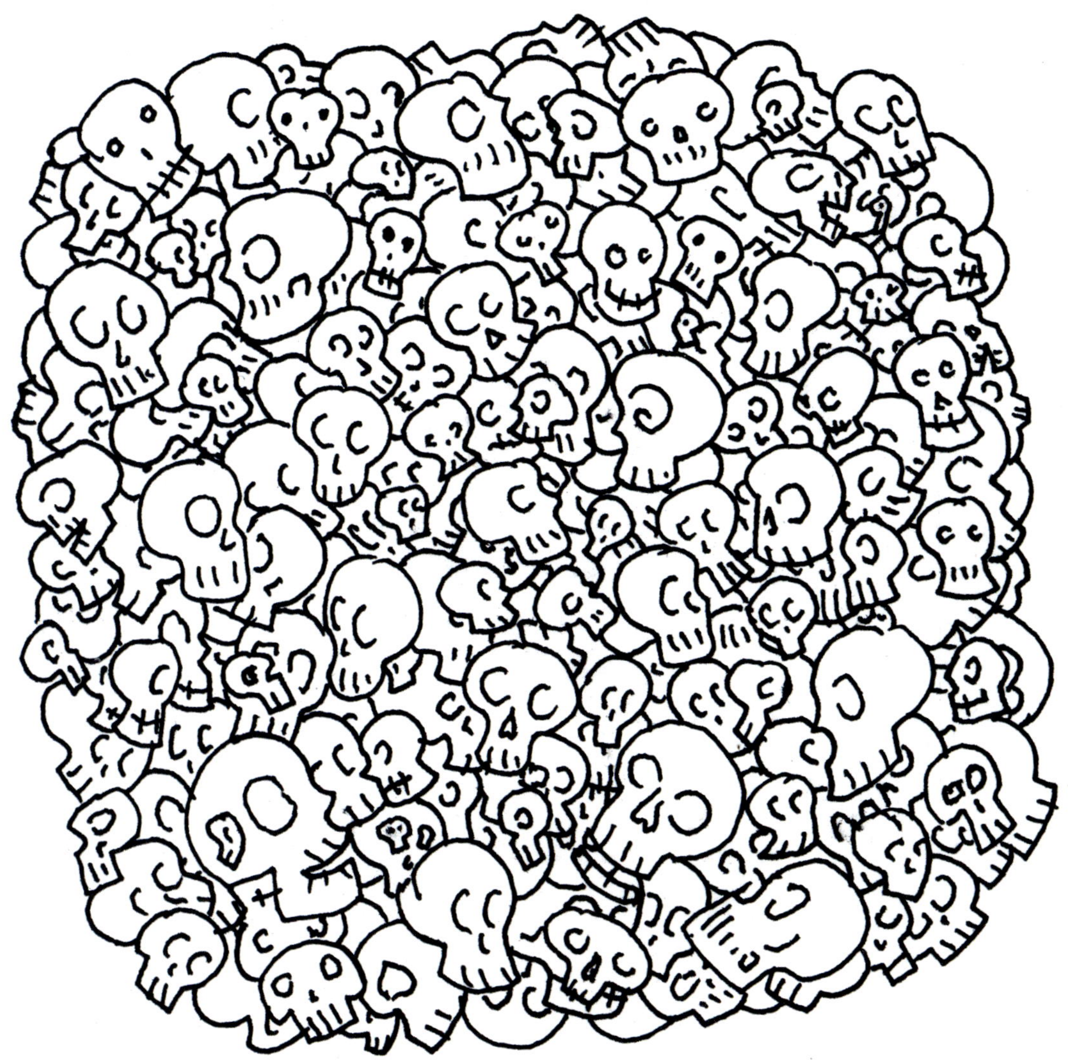

45 Minutes of Cartoon Skulls
Ink on paper, 5½ x 5½ inches (14 x 14cm).
(Continuously drawn for 45 minutes)

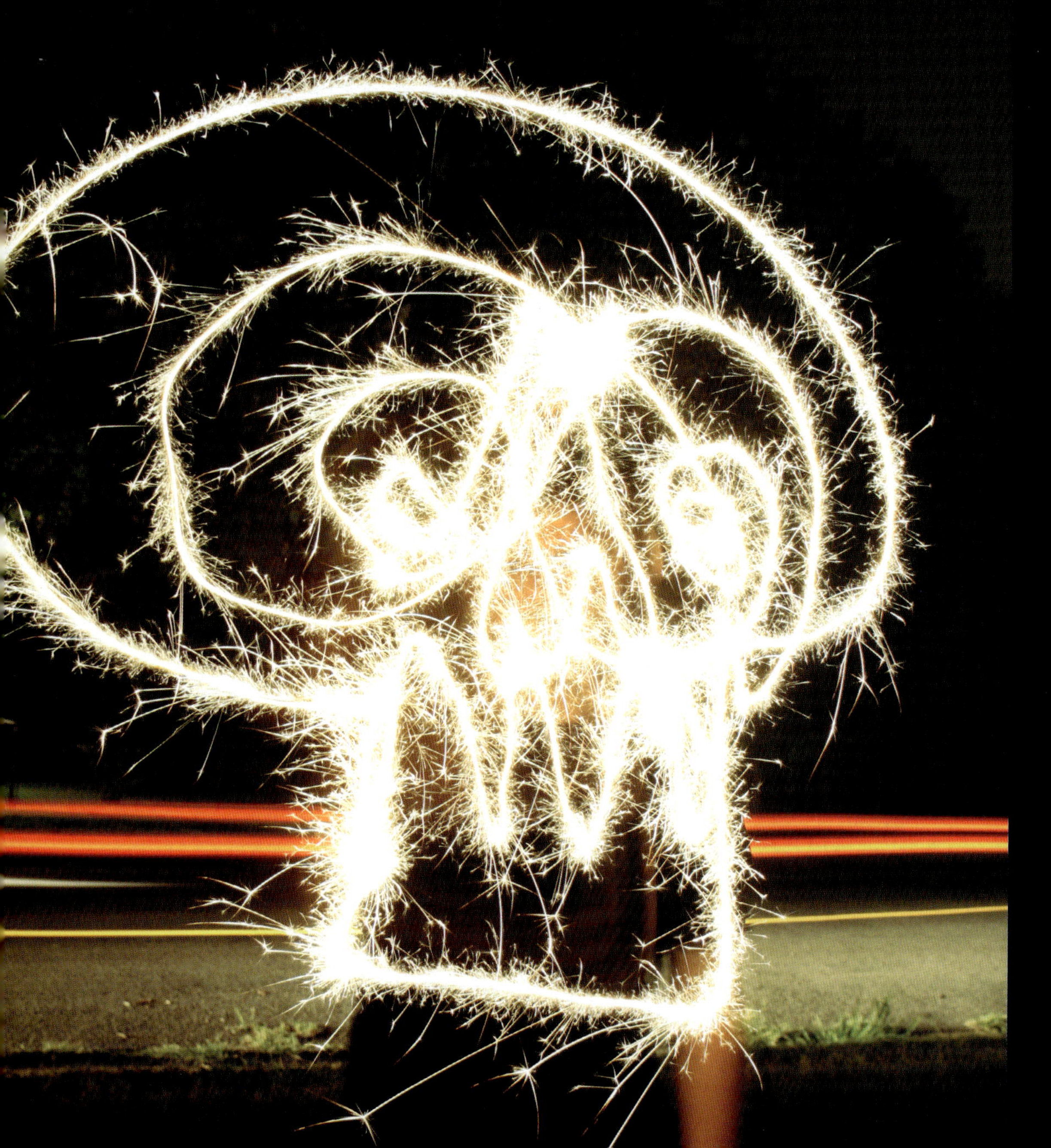

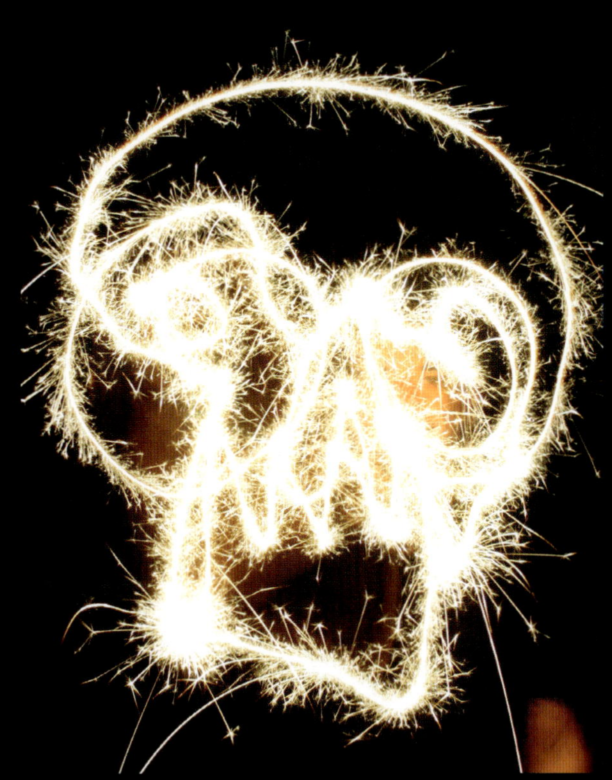

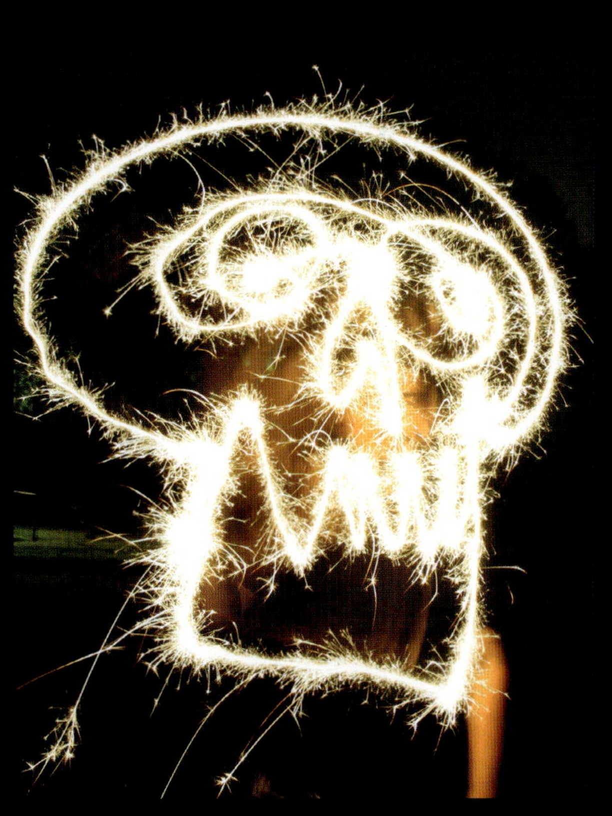

Sparkler Skull Long exposure sparkler drawing. (I did 22 versions of this one.)

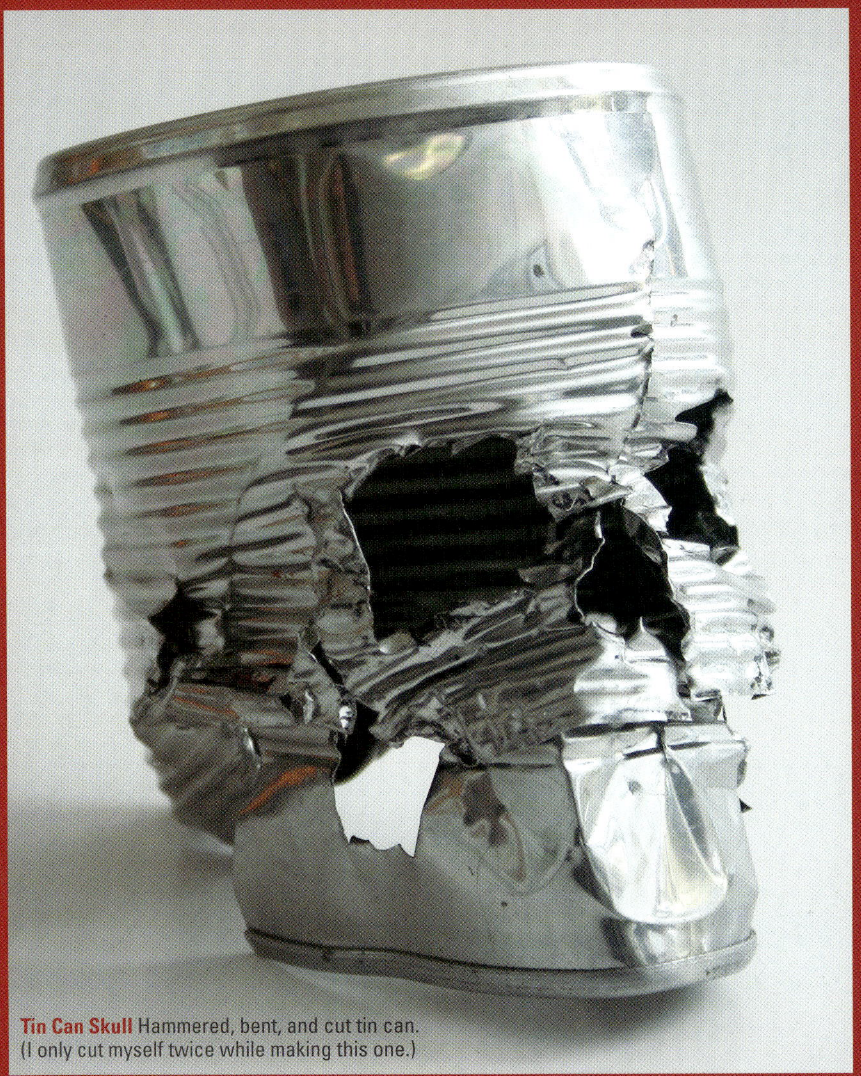

Tin Can Skull Hammered, bent, and cut tin can. (I only cut myself twice while making this one.)

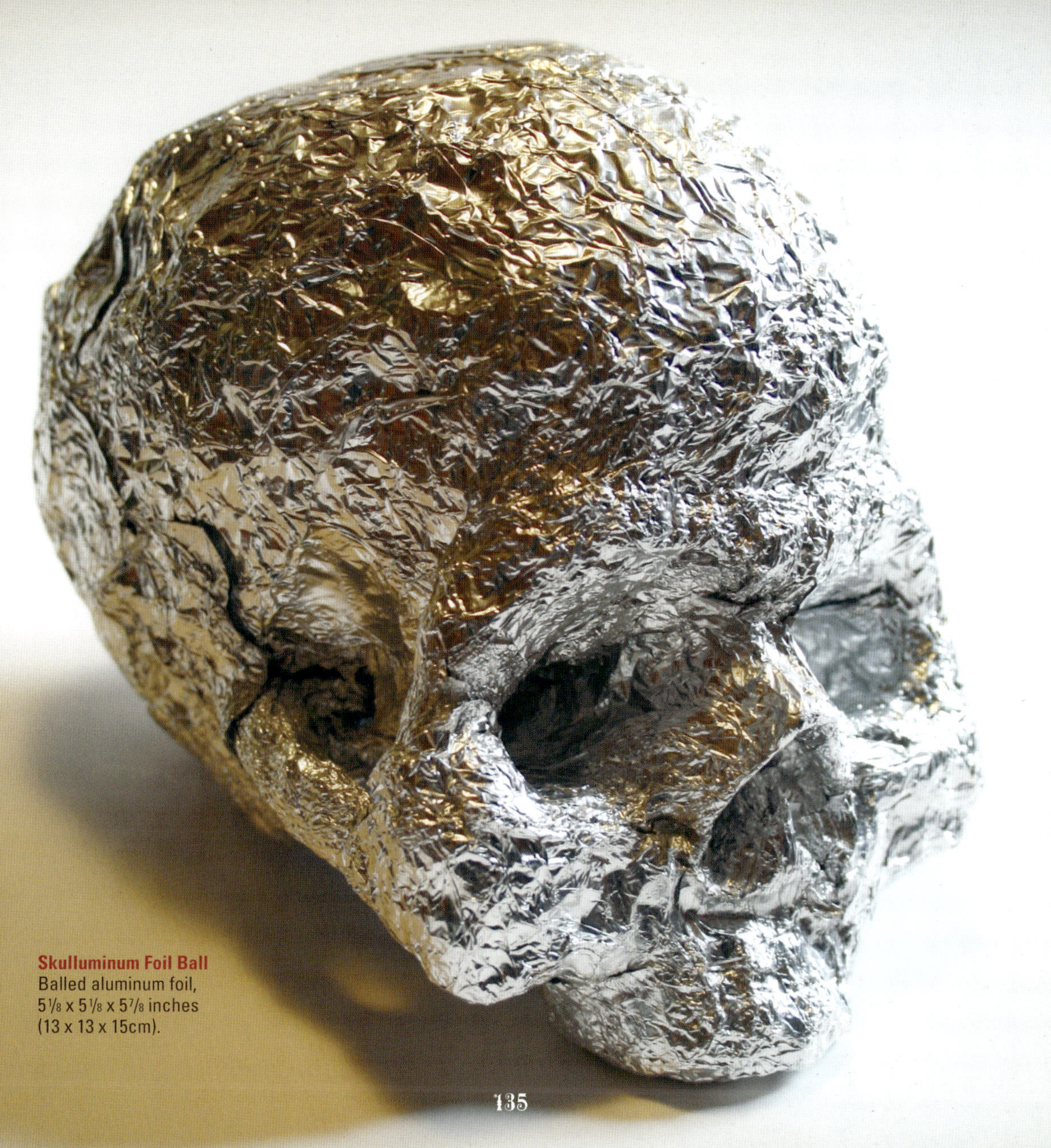

Skulluminum Foil Ball
Balled aluminum foil,
5 1/8 x 5 1/8 x 5 7/8 inches
(13 x 13 x 15cm).

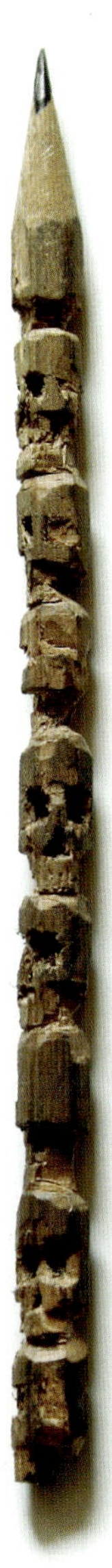

Totem Pencil Skull
Carved pencil
(two views)

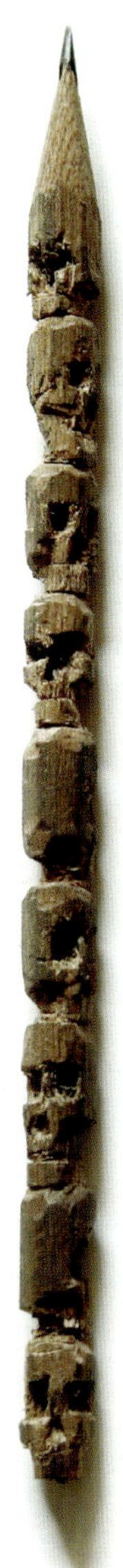

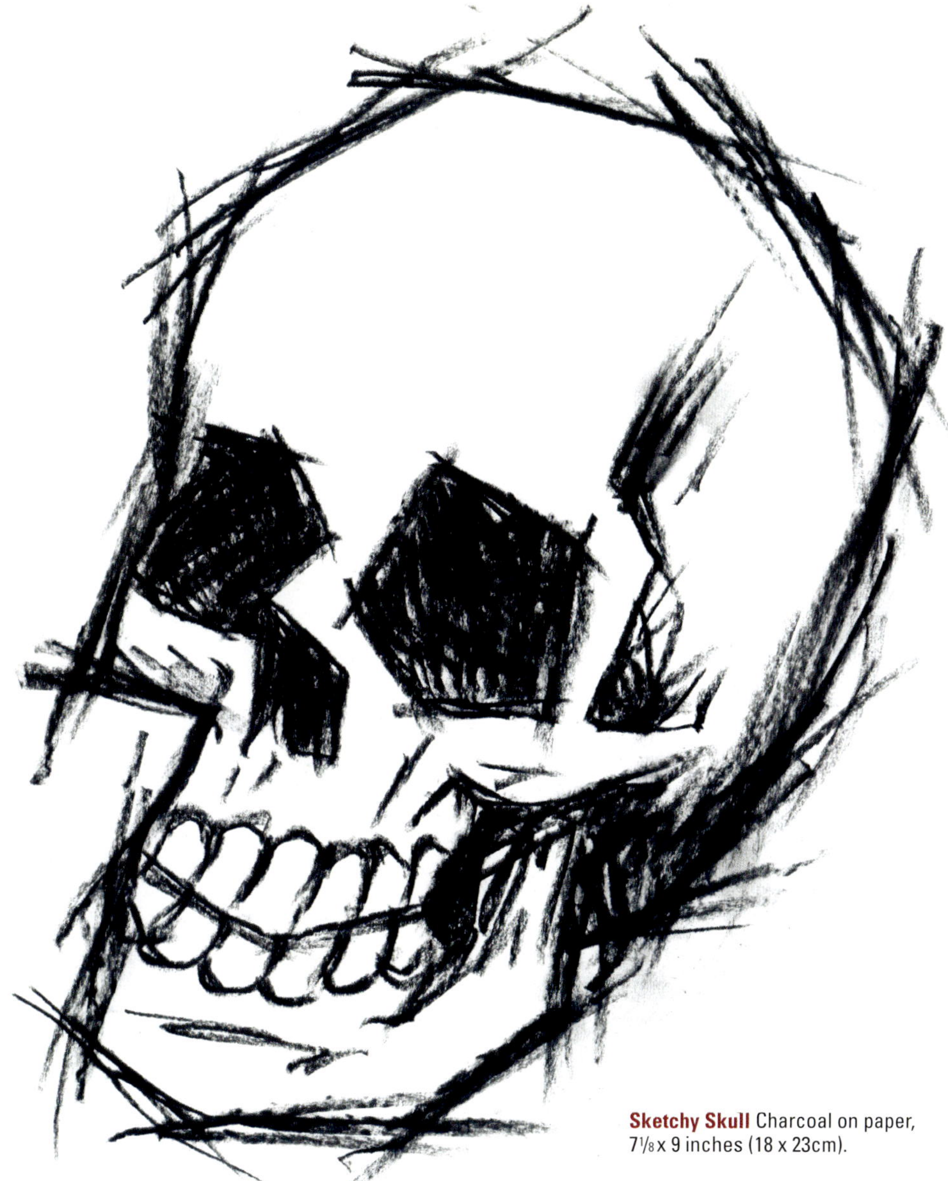

Sketchy Skull Charcoal on paper, 7 1/8 x 9 inches (18 x 23cm).

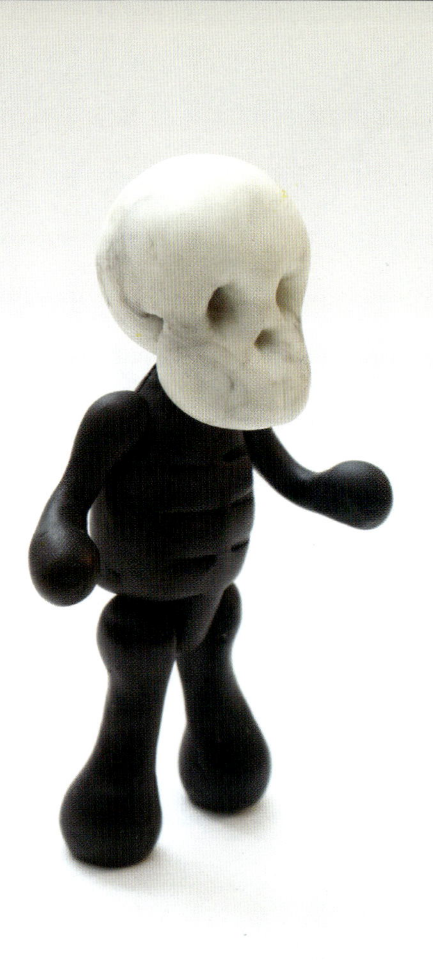

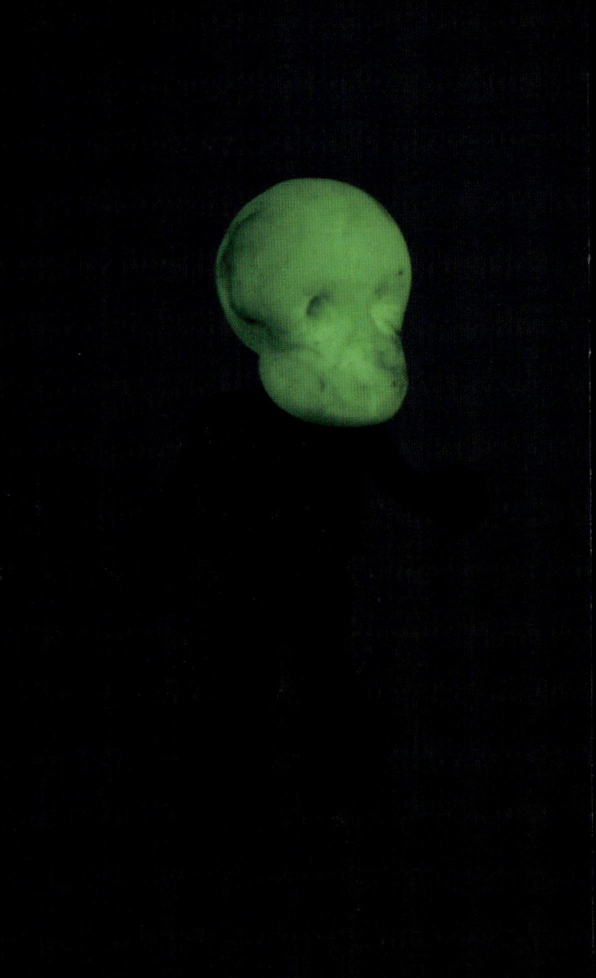

Action Figure Skull
Sculpted modeling compound, 3⅞ inches (10cm) tall. (It glows in the dark!)

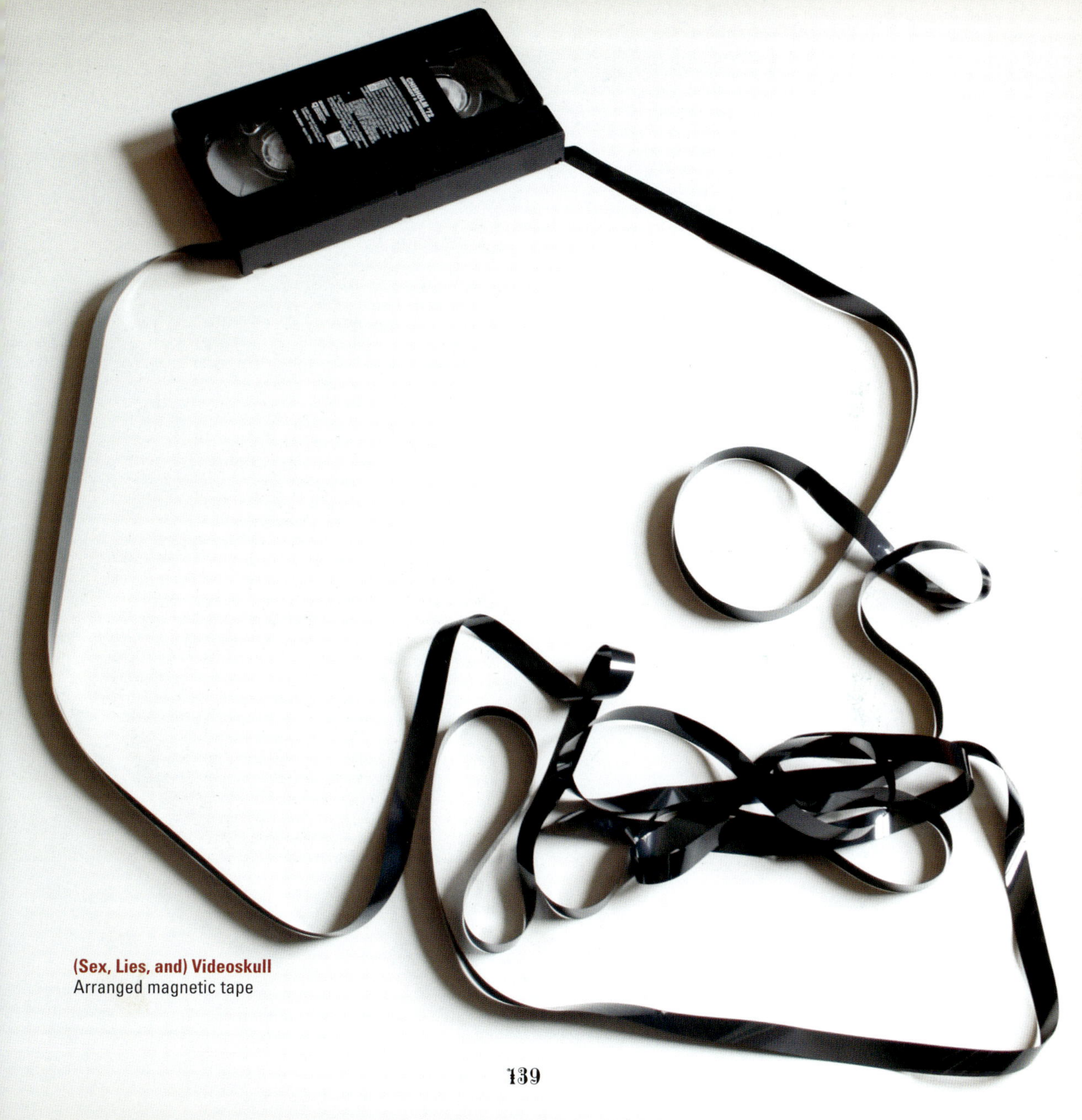

(Sex, Lies, and) Videoskull
Arranged magnetic tape

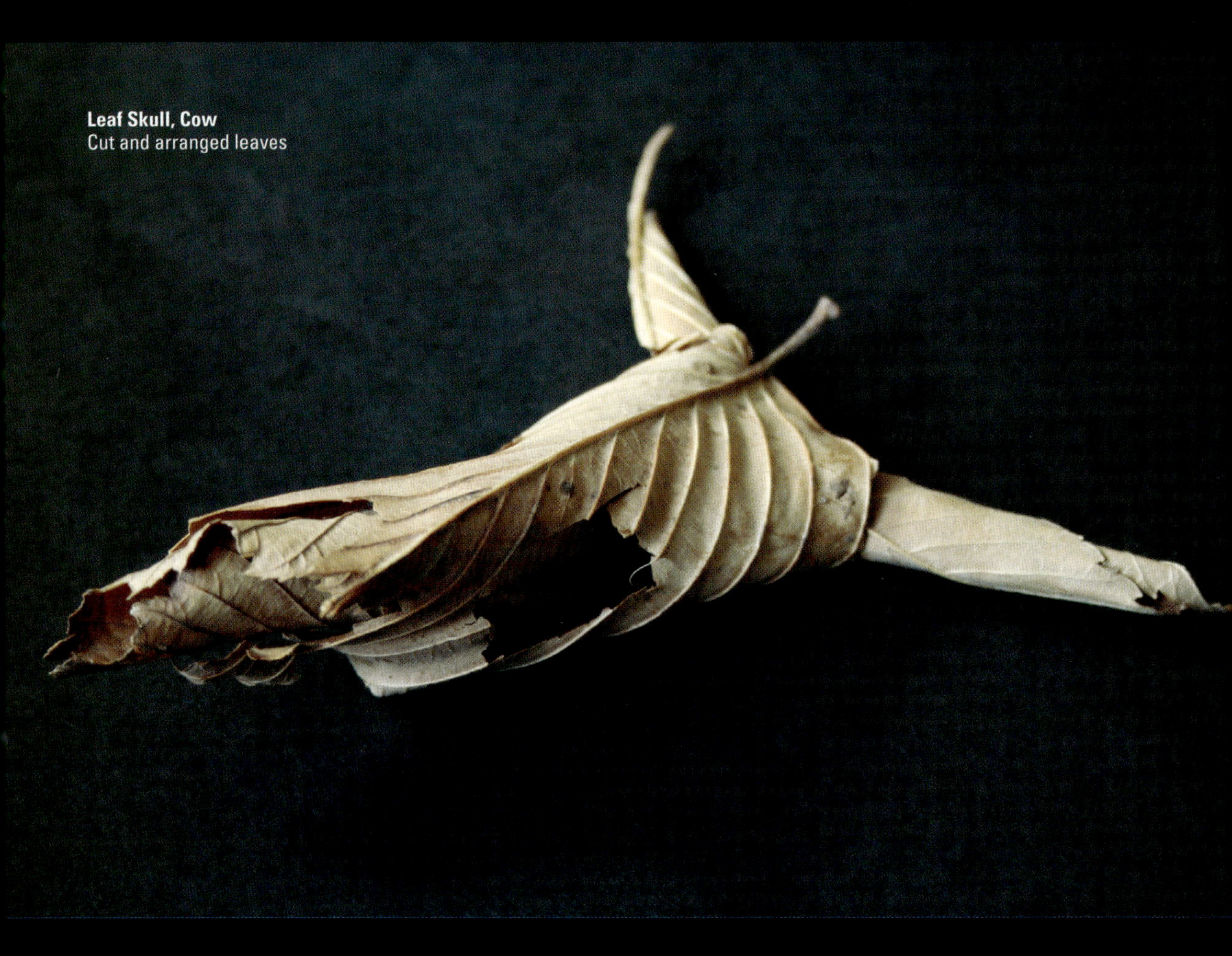

Leaf Skull, Cow
Cut and arranged leaves

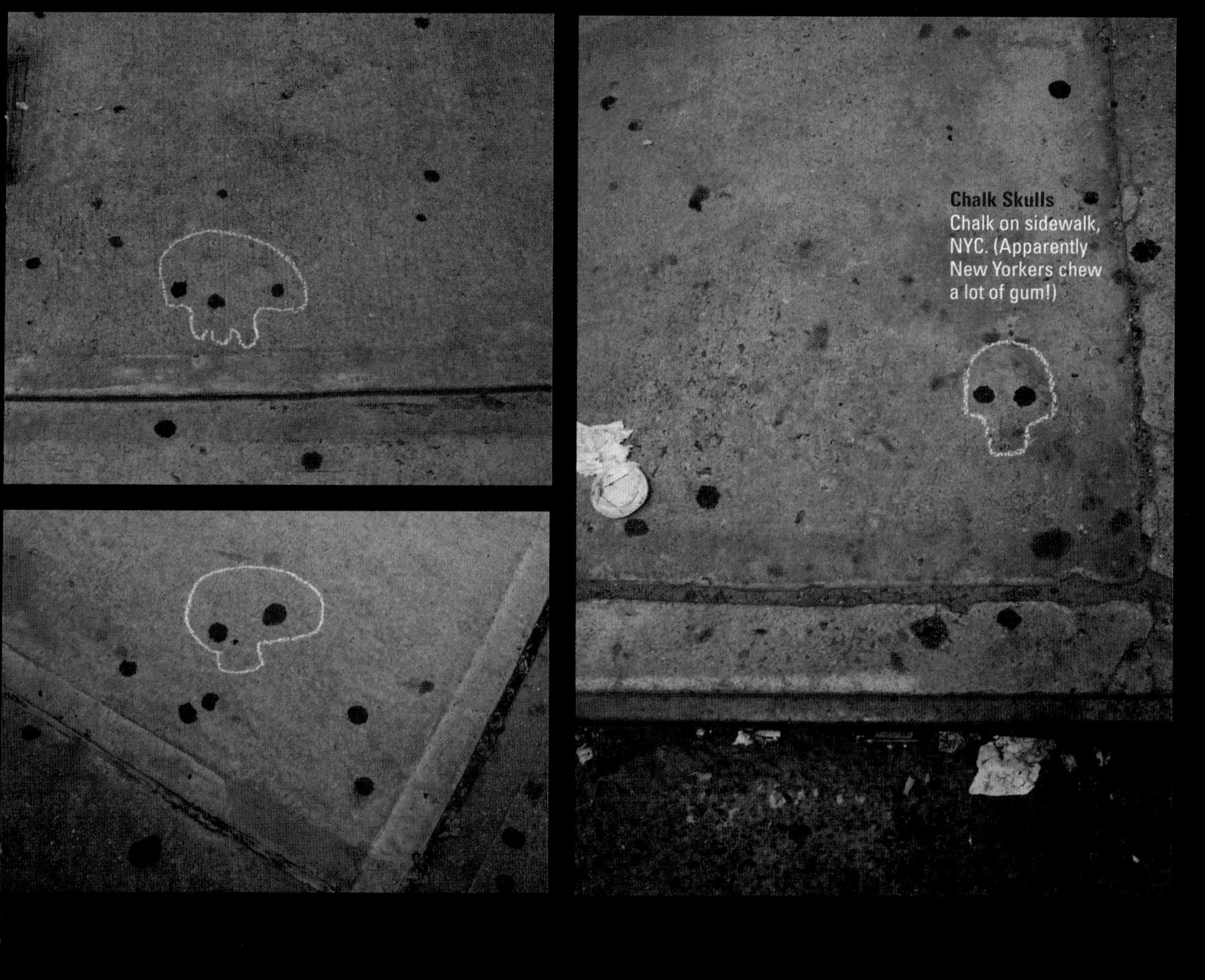

Chalk Skulls
Chalk on sidewalk, NYC. (Apparently New Yorkers chew a lot of gum!)

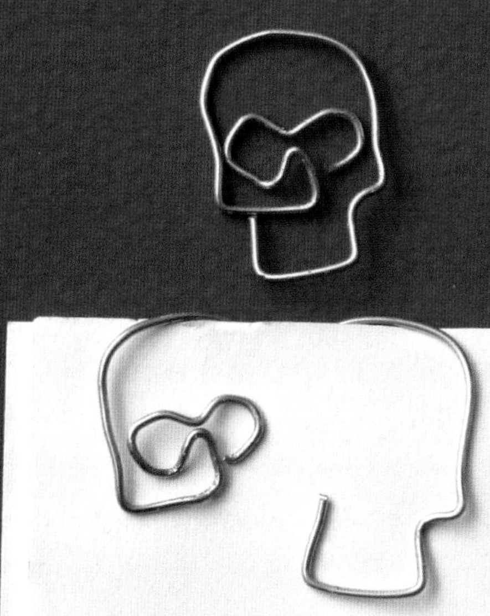

Skull Clips Bent 20-gauge steel wire, each 1 x 5/8 inches (2.5 x 1.5cm). (They are functional, though remarkably time intensive to make.)

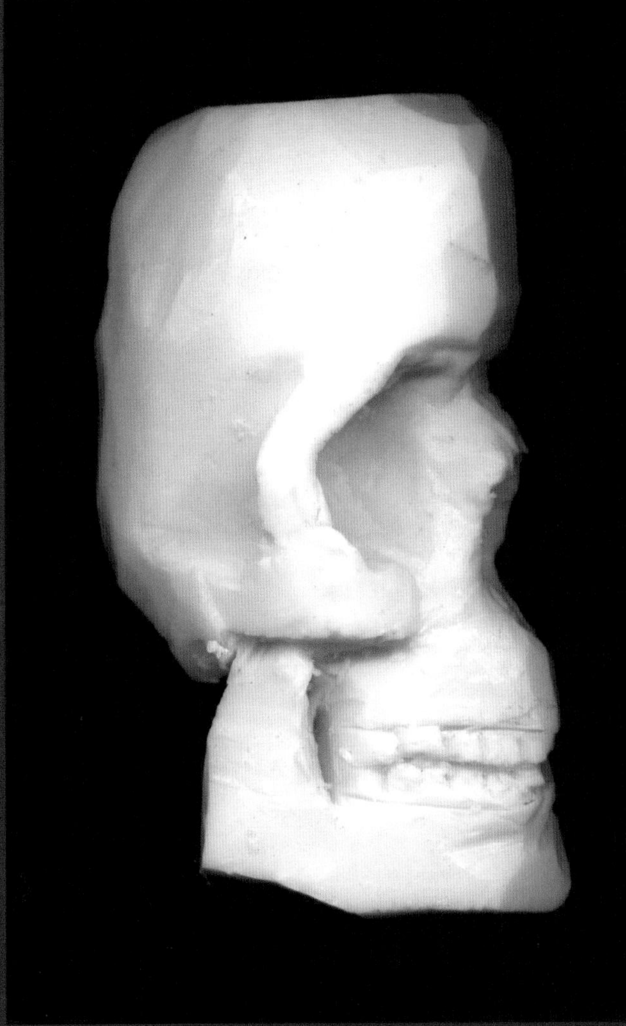

Bar of Skull Carved used bar of soap. (I still have a scar on my finger from the last time I tried carving soap as a teenager.)

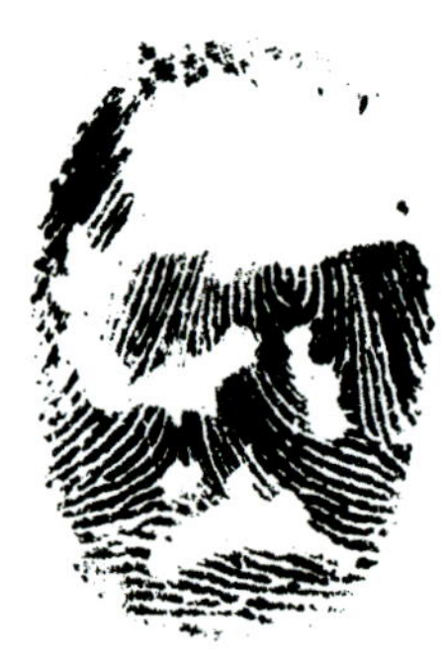

Fingerprint Skull
Ink on finger
monoprint

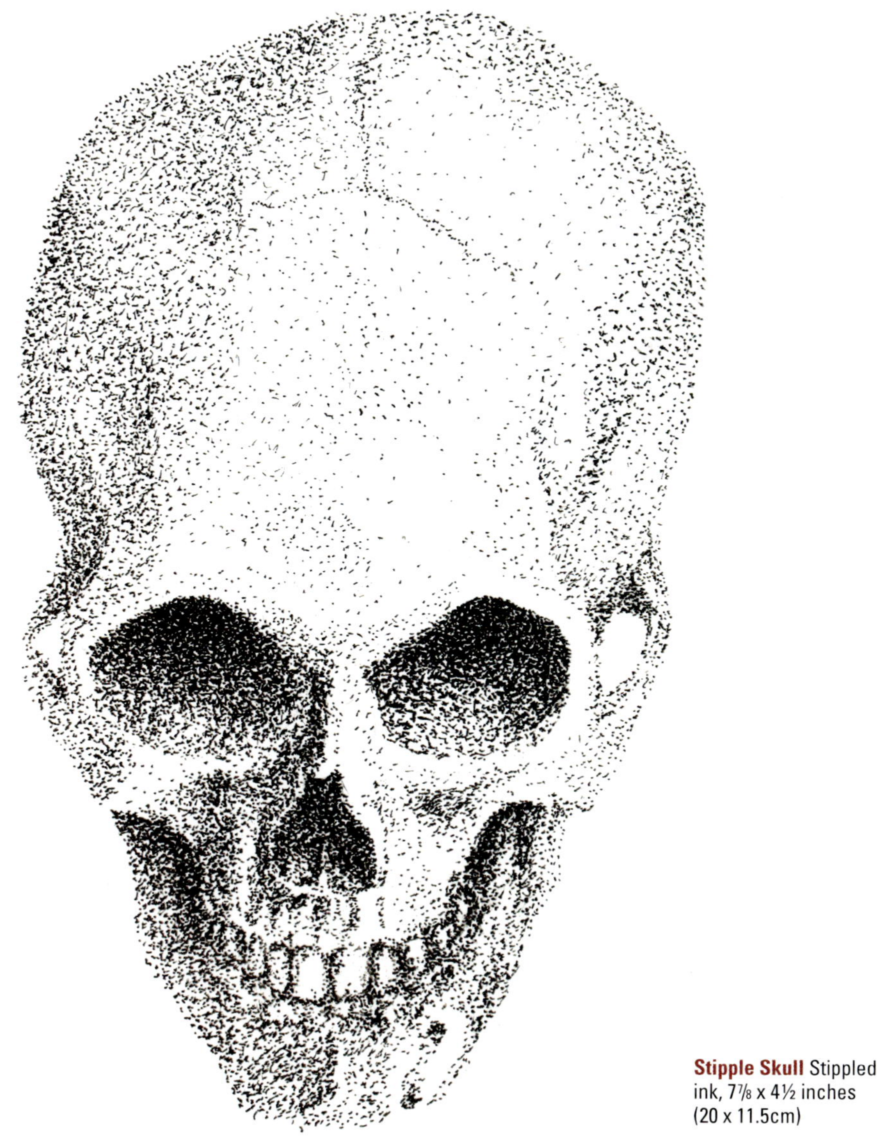

Stipple Skull Stippled ink, 7⅞ x 4½ inches (20 x 11.5cm)

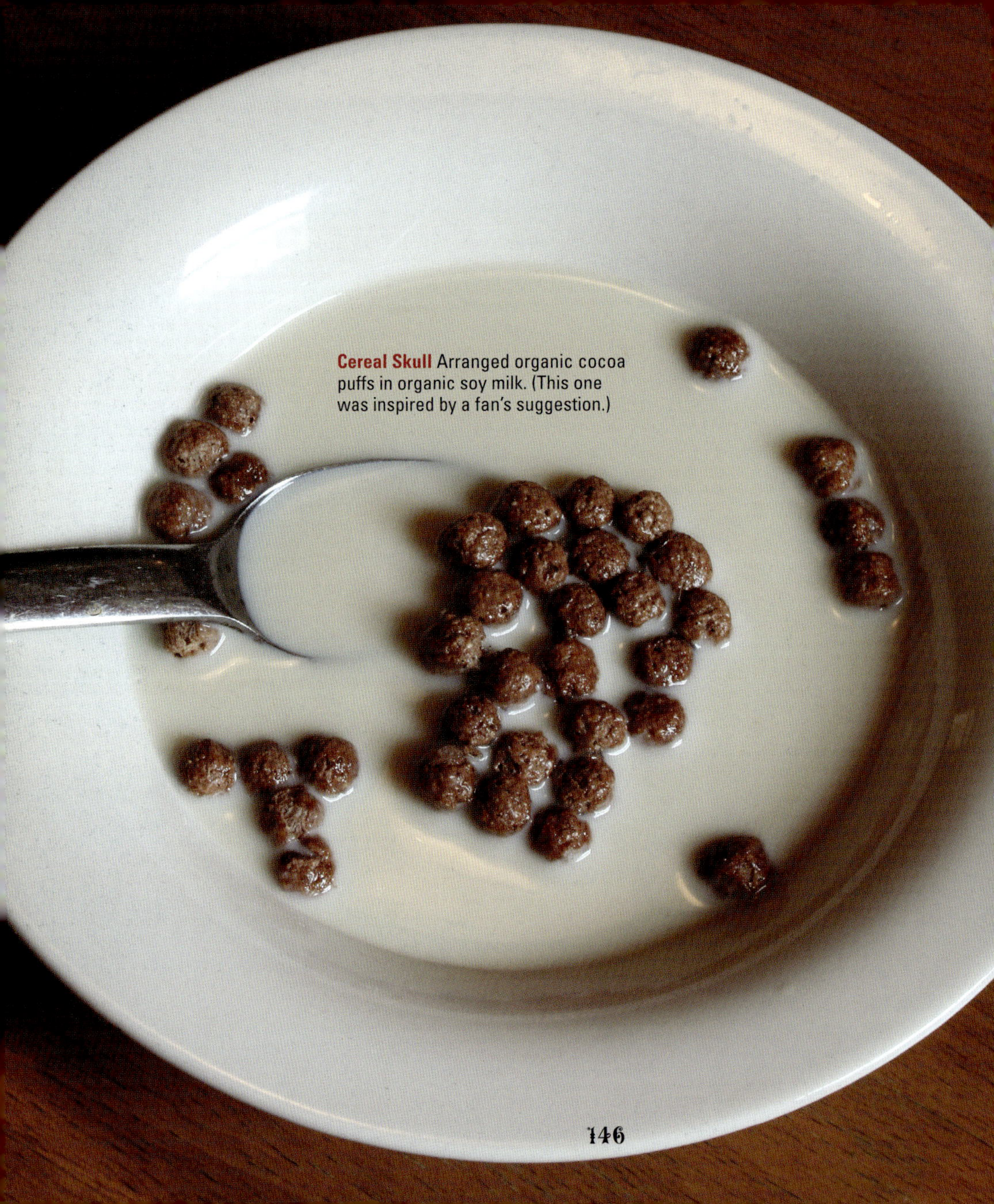

Cereal Skull Arranged organic cocoa puffs in organic soy milk. (This one was inspired by a fan's suggestion.)

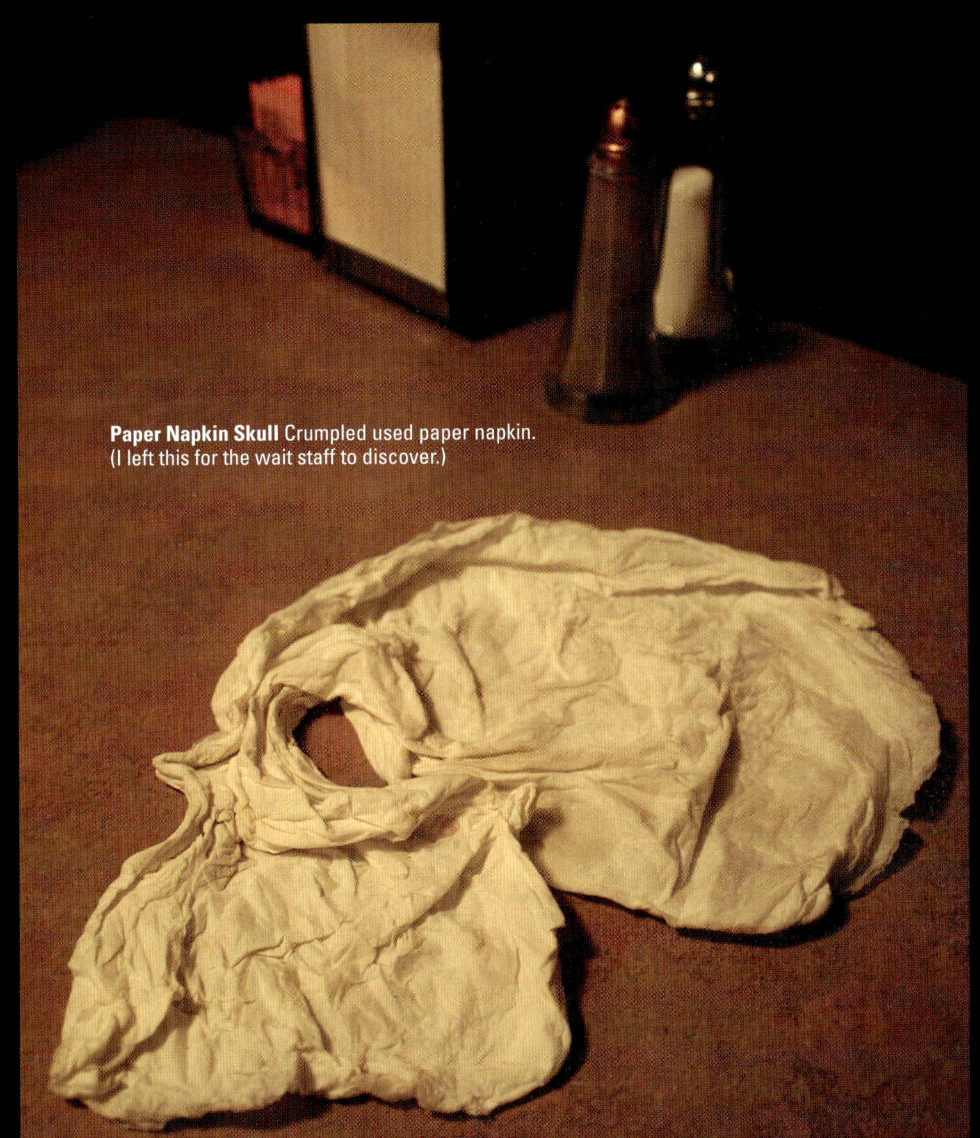

Paper Napkin Skull Crumpled used paper napkin.
(I left this for the wait staff to discover.)

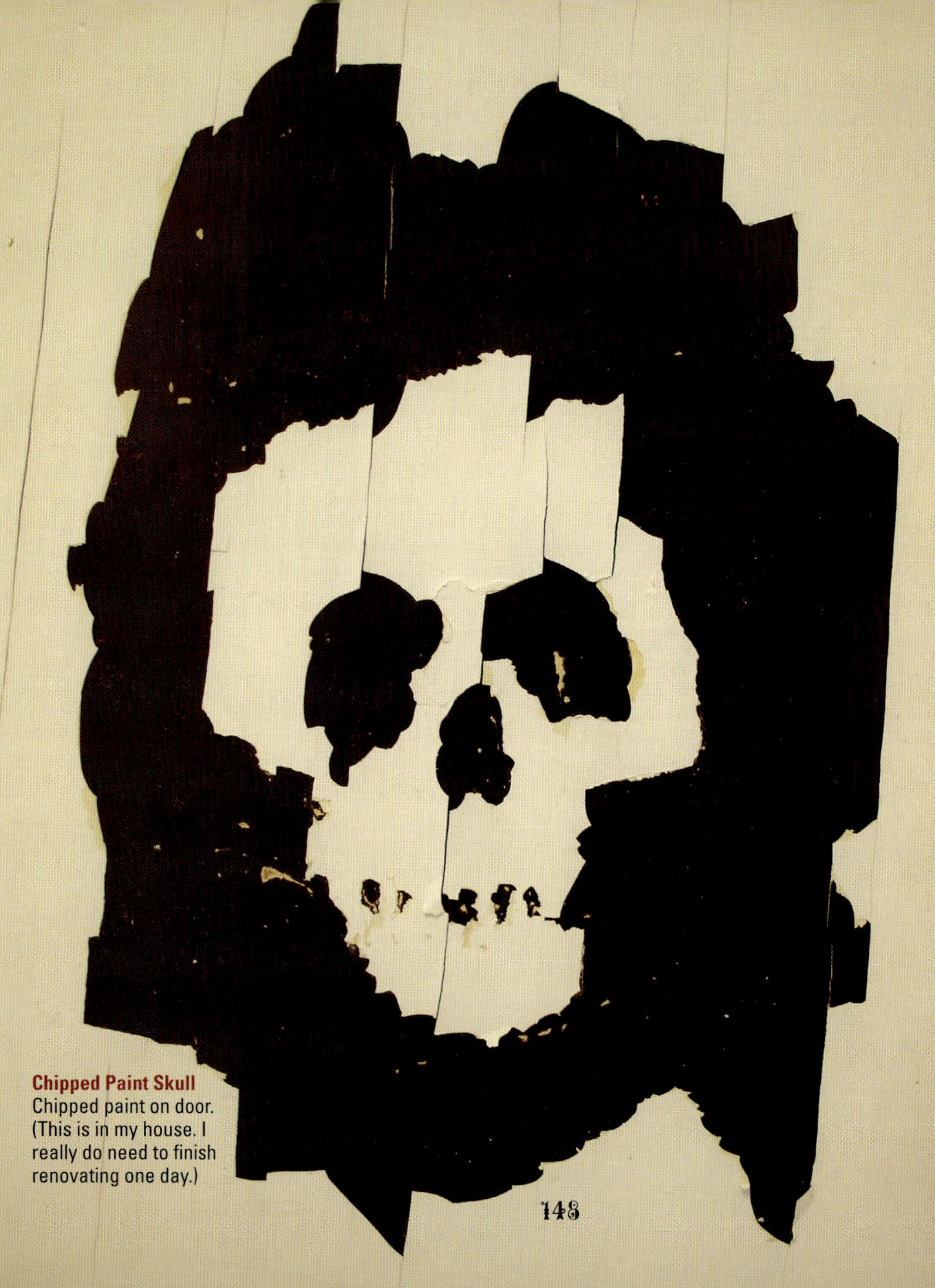

Chipped Paint Skull
Chipped paint on door. (This is in my house. I really do need to finish renovating one day.)

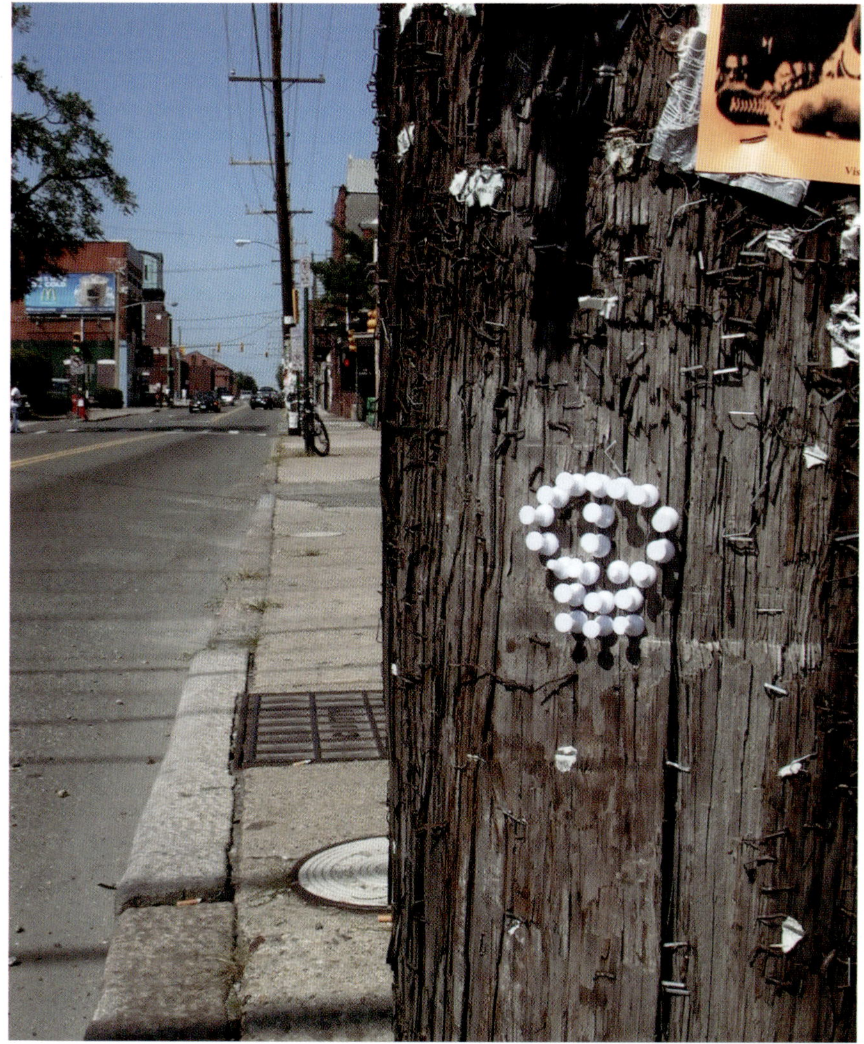

Pushpin Skull Pushpins hammered into electrical pole. (This actually stayed in place for several months.)

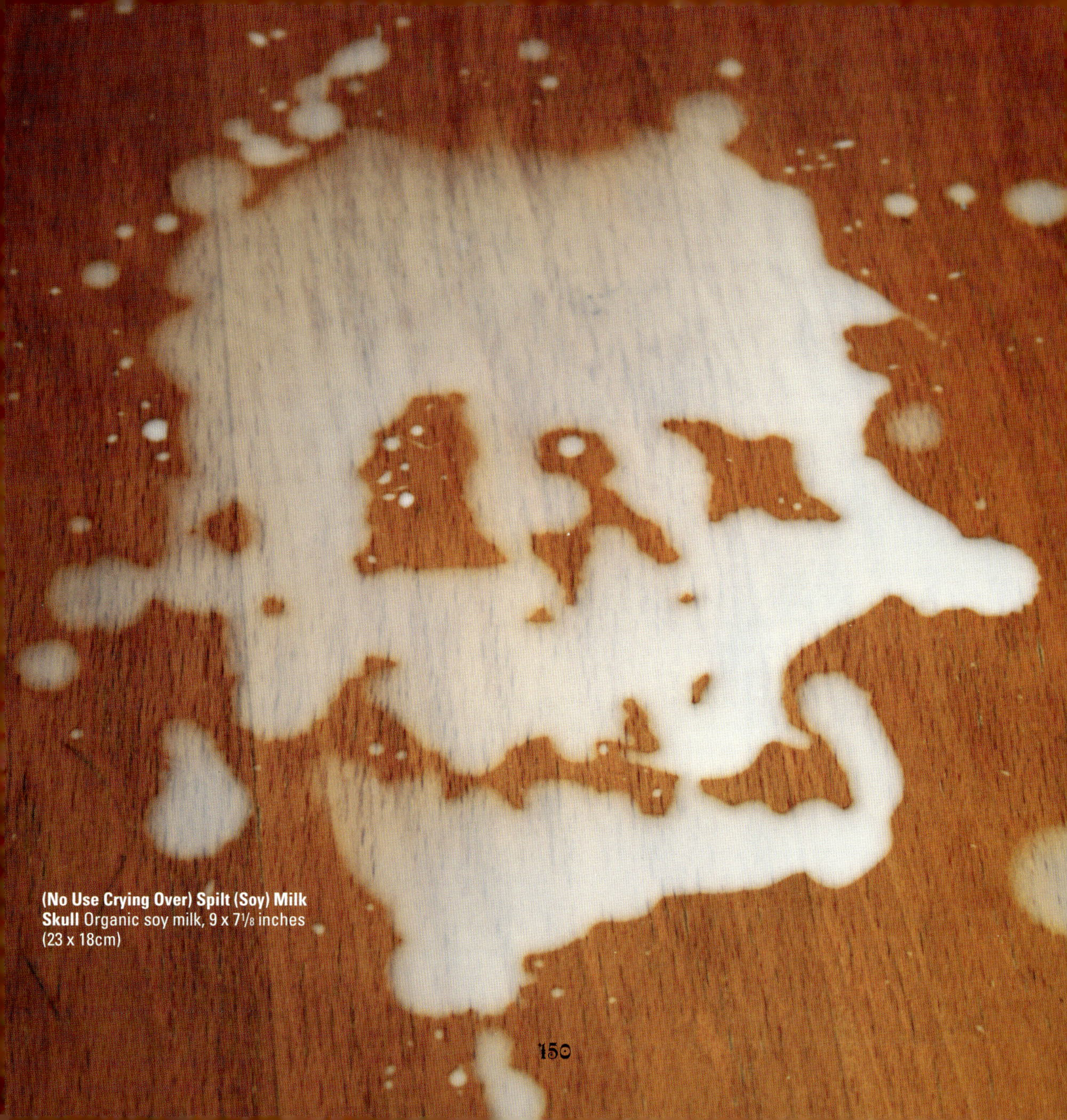

(No Use Crying Over) Spilt (Soy) Milk Skull Organic soy milk, 9 x 7⅛ inches (23 x 18cm)

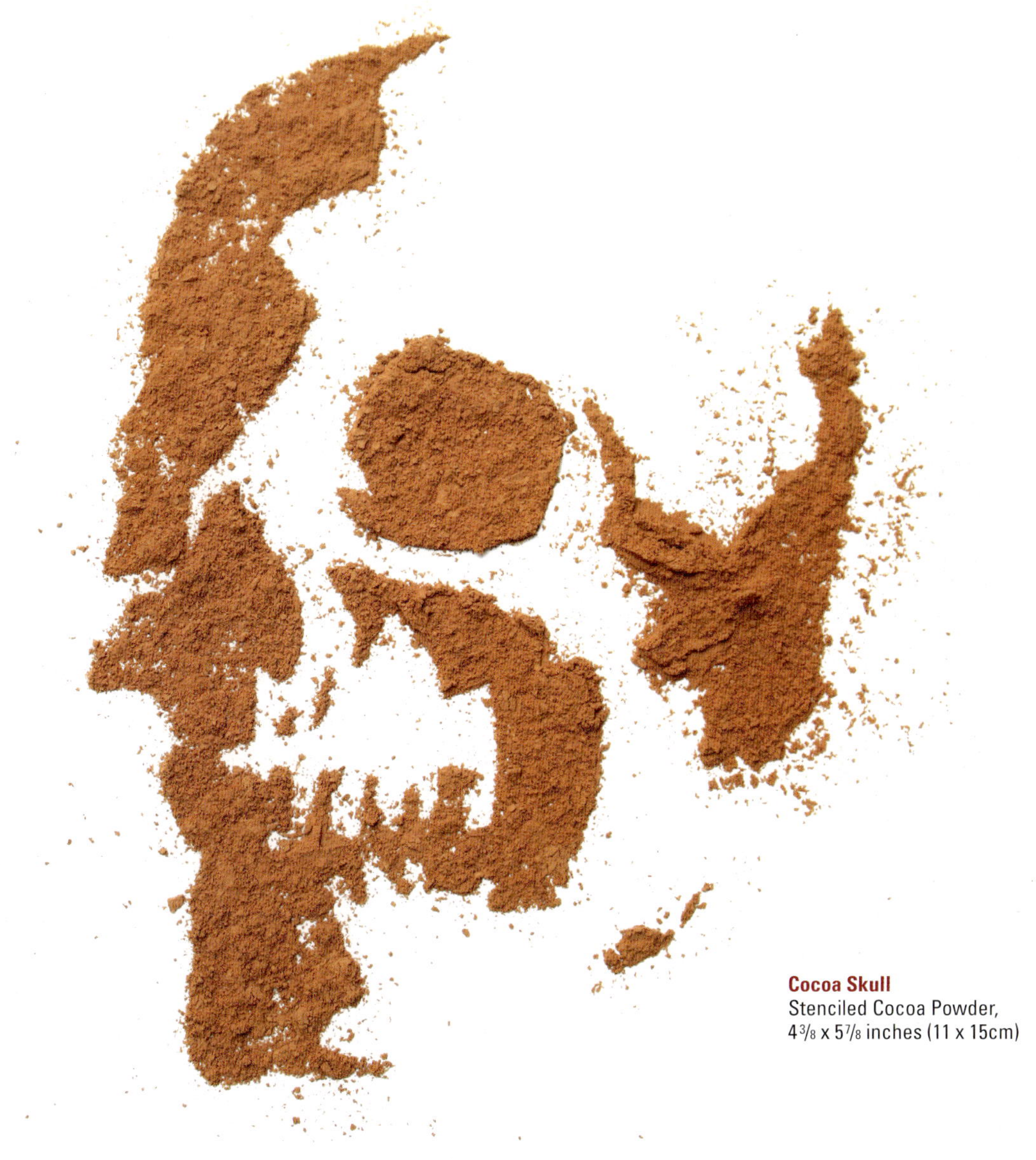

Cocoa Skull
Stenciled Cocoa Powder,
4 3/8 x 5 7/8 inches (11 x 15cm)

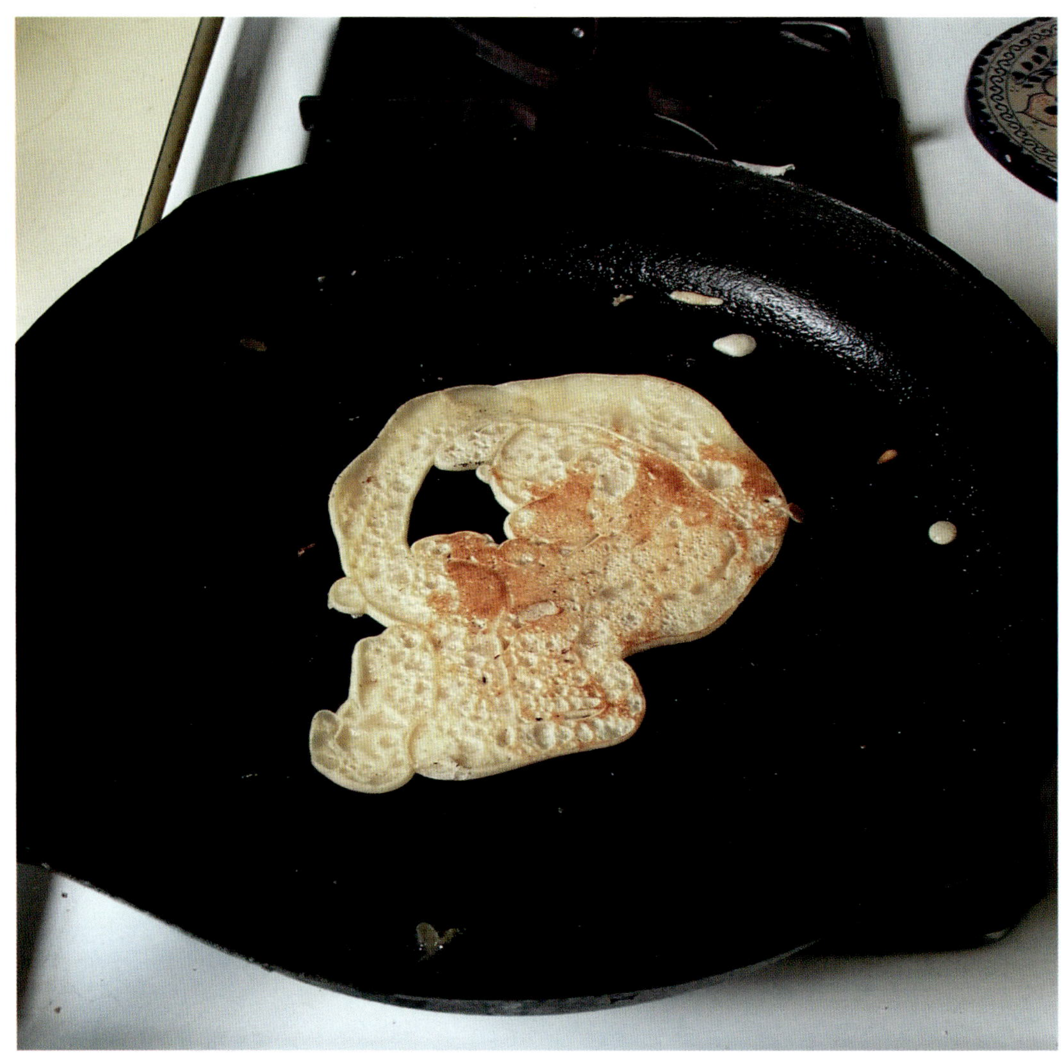

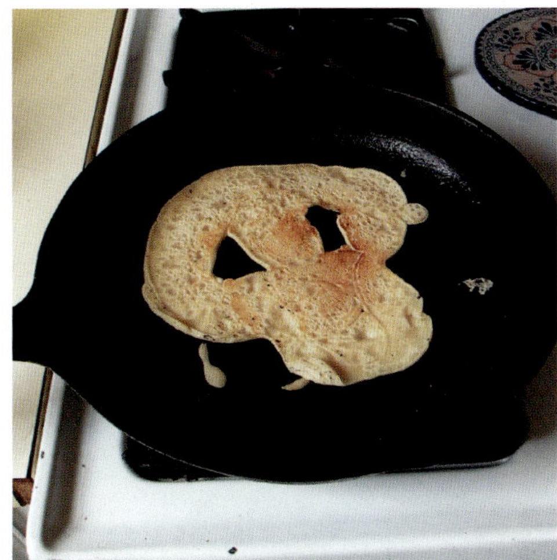

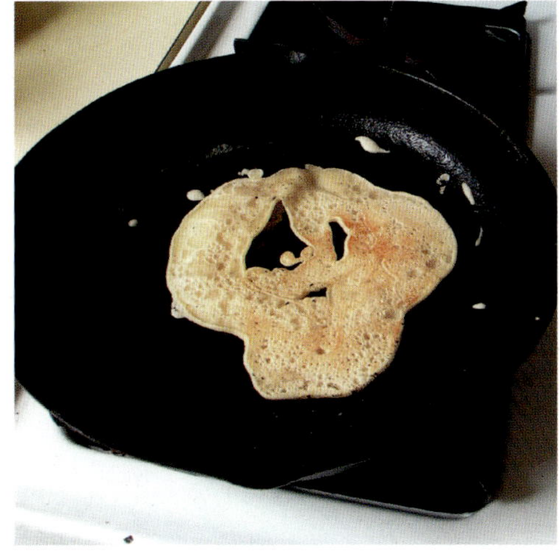

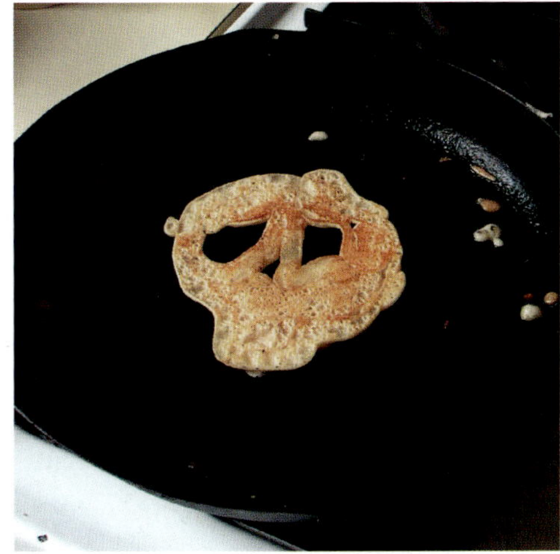

Skull Pancakes Vegan pancake batter. (They were delicious!)

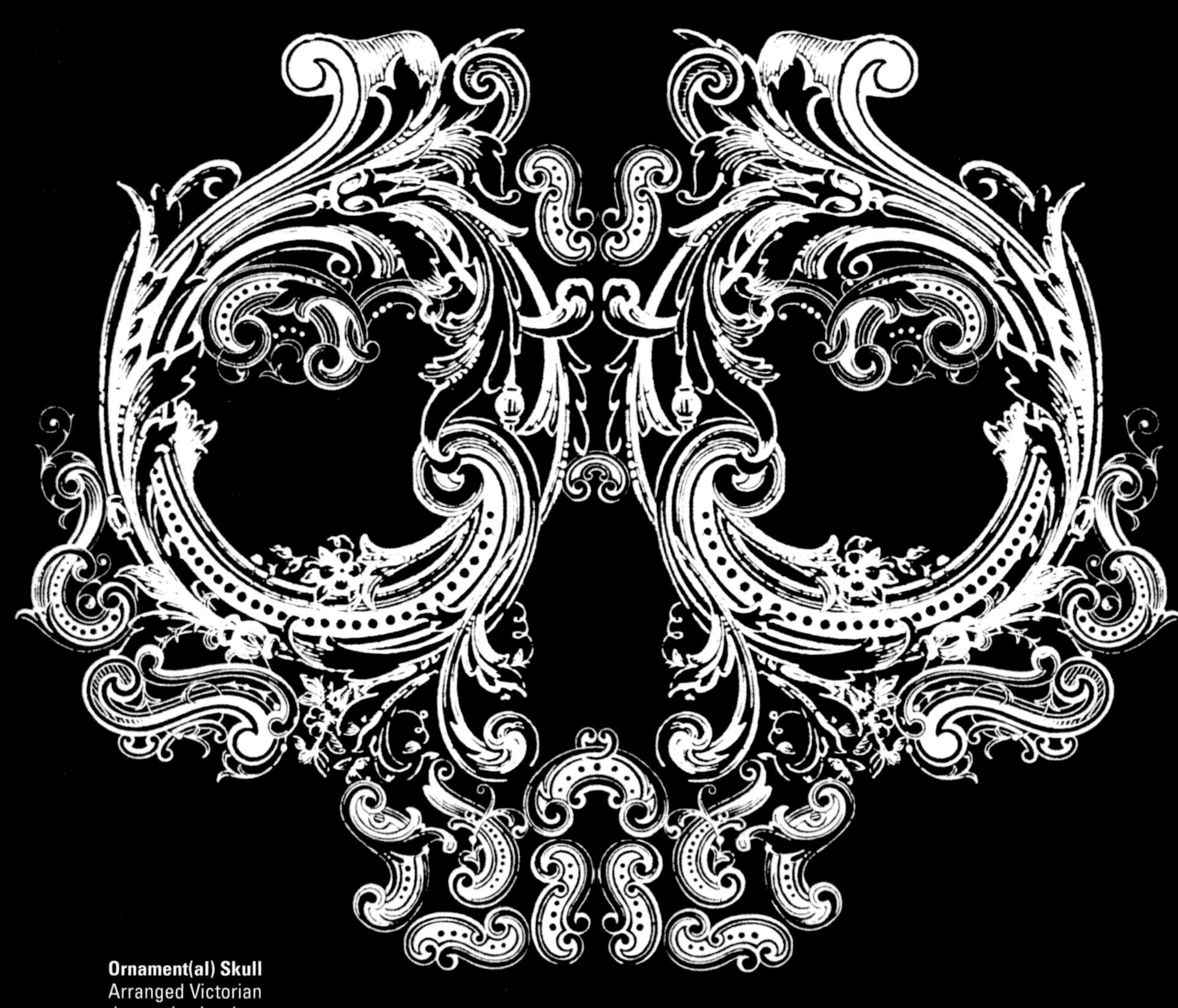

Ornament(al) Skull
Arranged Victorian decorative border elements

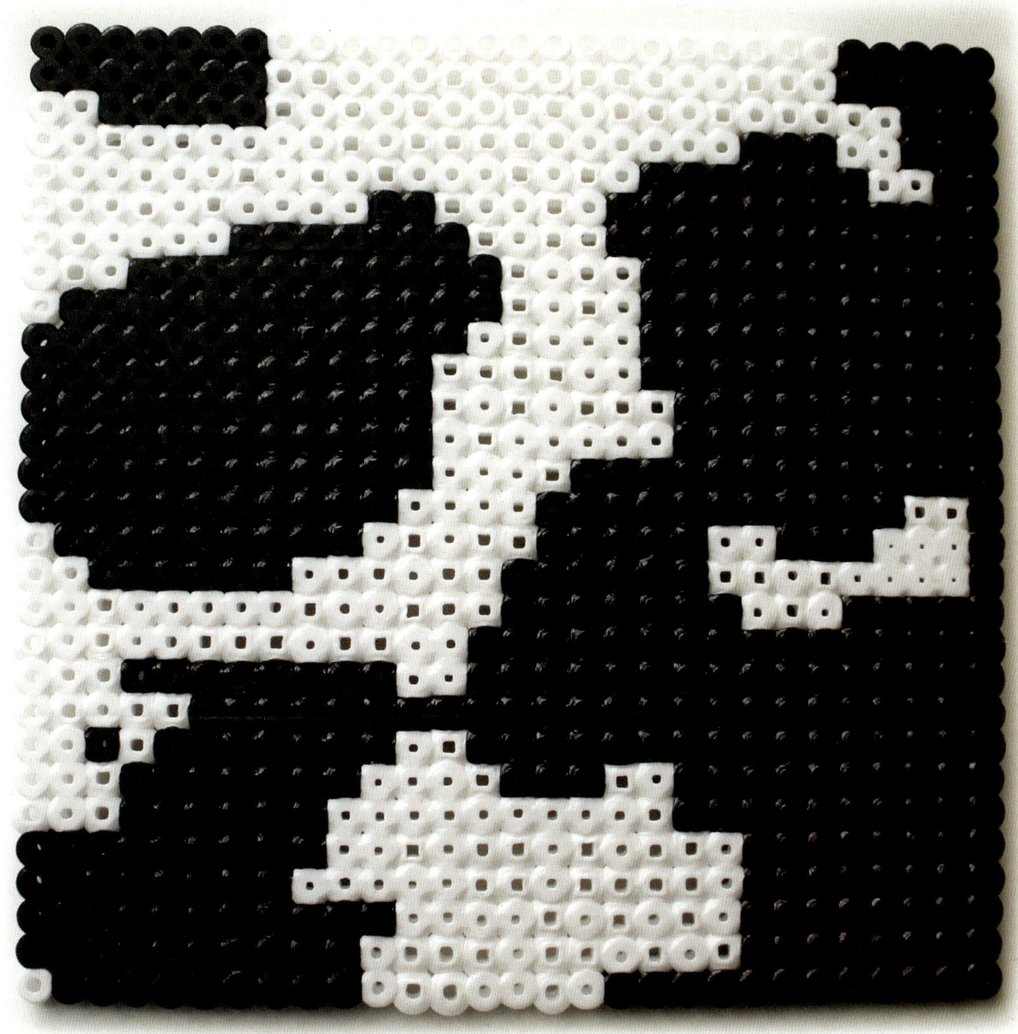

Fused Bead Skull Fused plastic beads,
5½ x 5½ inches (14 x 14cm)

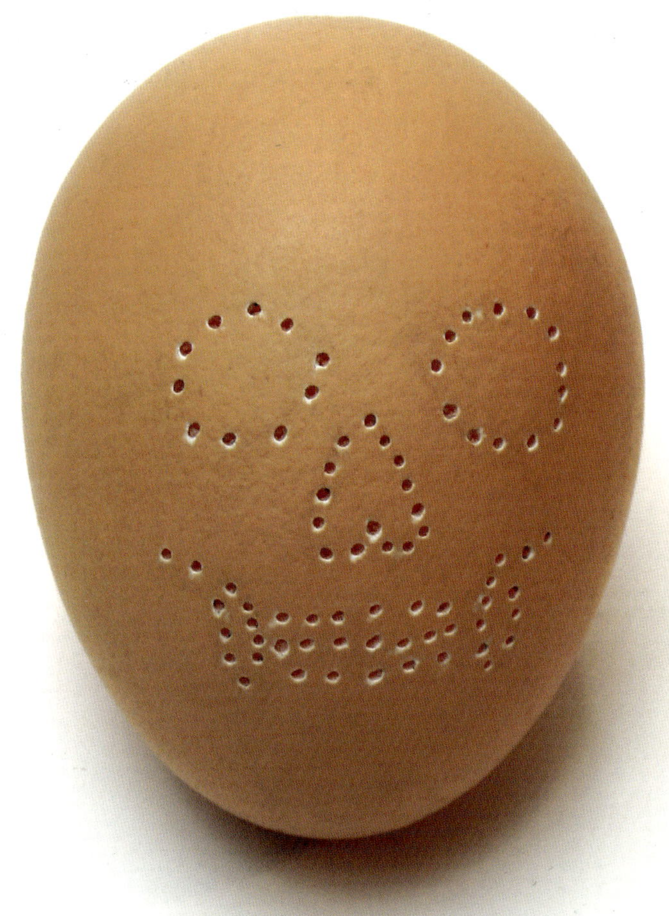

Egg Skull Drilled, local organic free-range egg. (Amazingly, I only broke two eggs and one drill bit before I got this to work.)

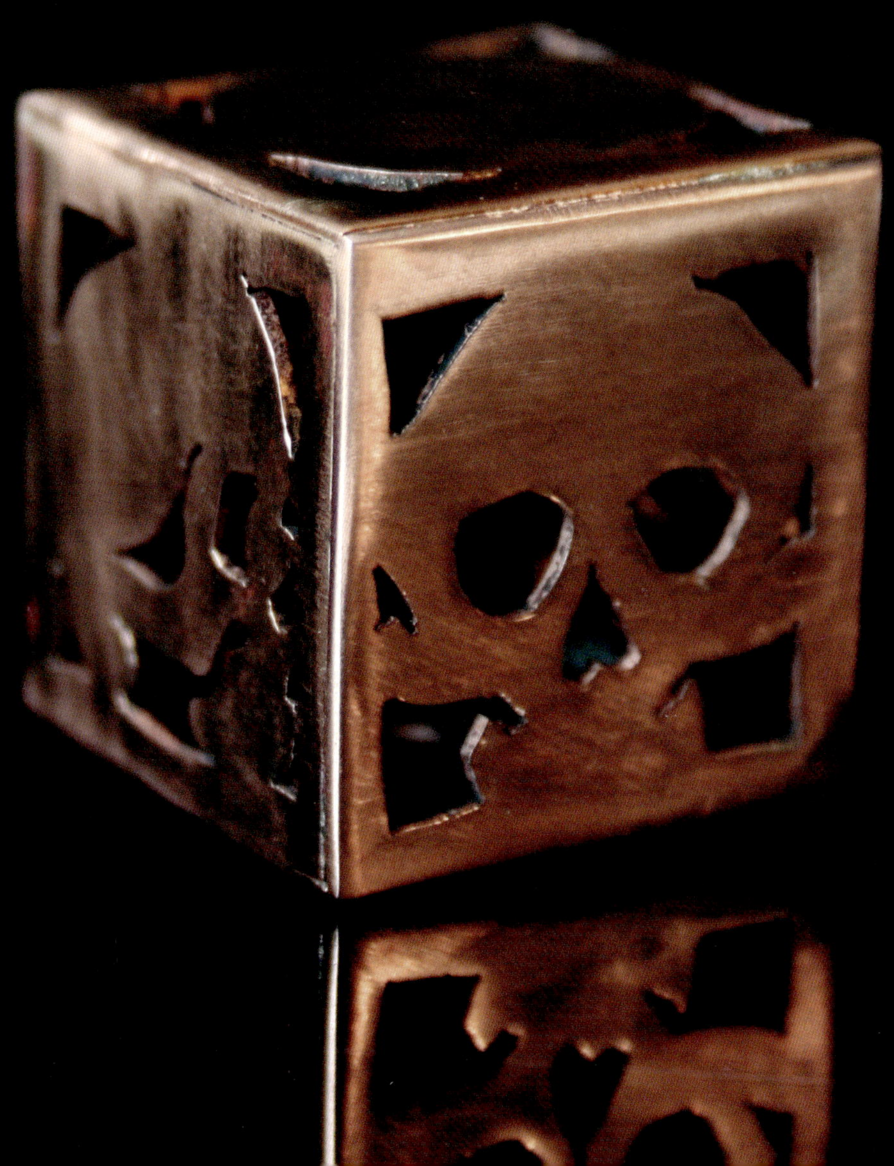

Copper Cube Skull Cut and soldered 18-gauge copper sheet, 1 x 1 x 1 inch (2.5 x 2.5 x 2.5cm). (All six sides show different views of the skull.)

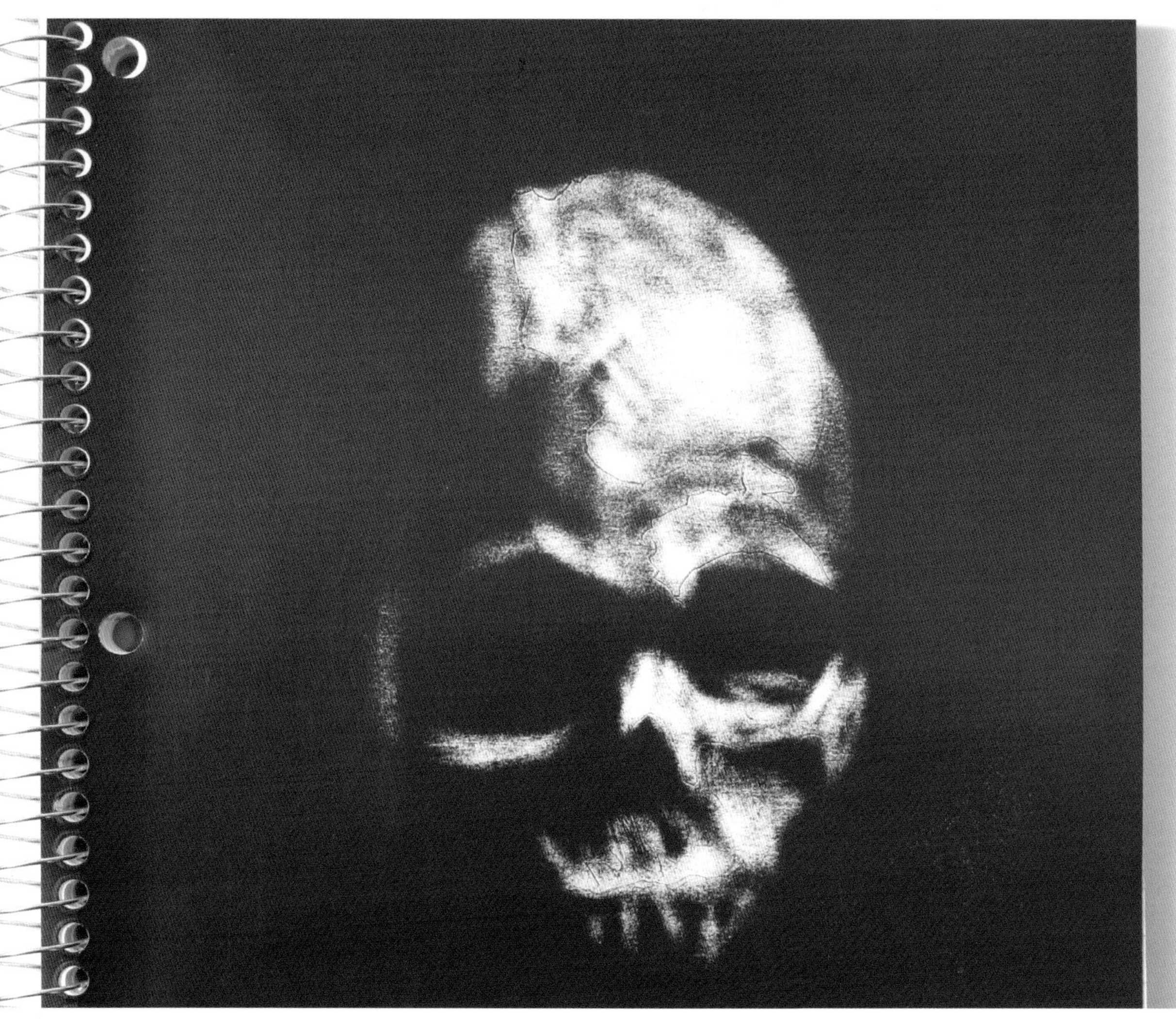

(Back to) Skull Notebook
Erasered notebook cover (I used to write my name on my notebooks when I was a kid using this same technique.)

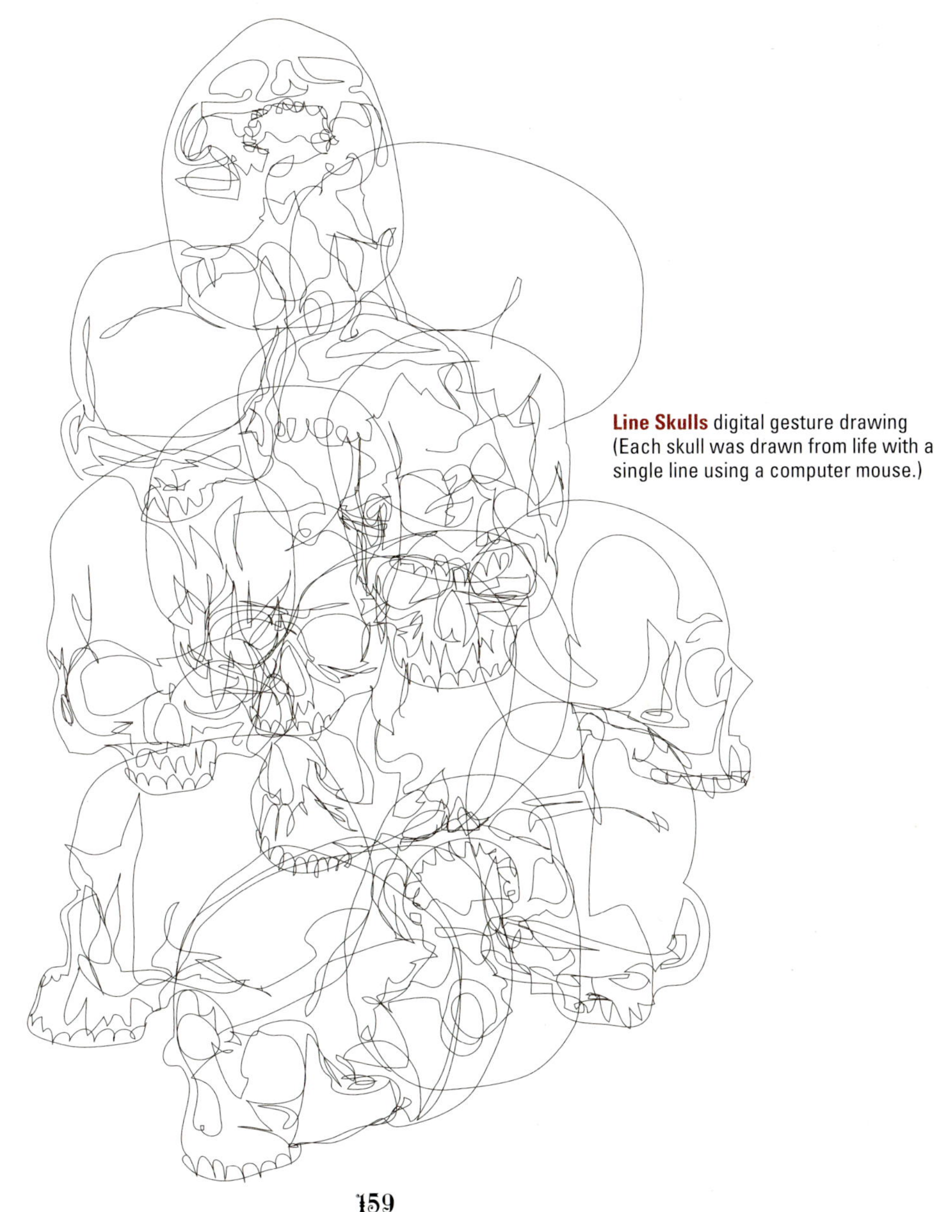

Line Skulls digital gesture drawing (Each skull was drawn from life with a single line using a computer mouse.)

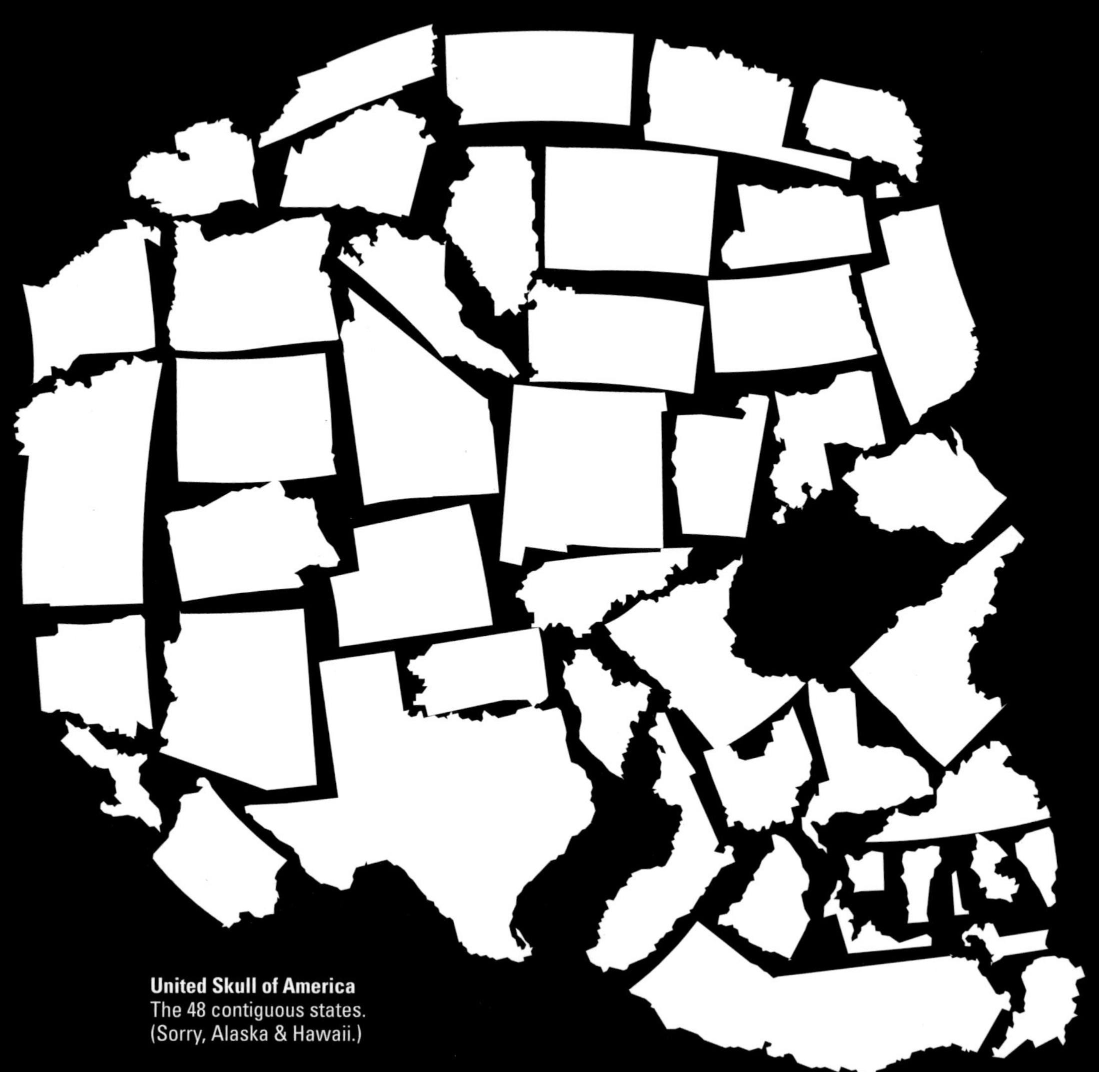

United Skull of America
The 48 contiguous states.
(Sorry, Alaska & Hawaii.)

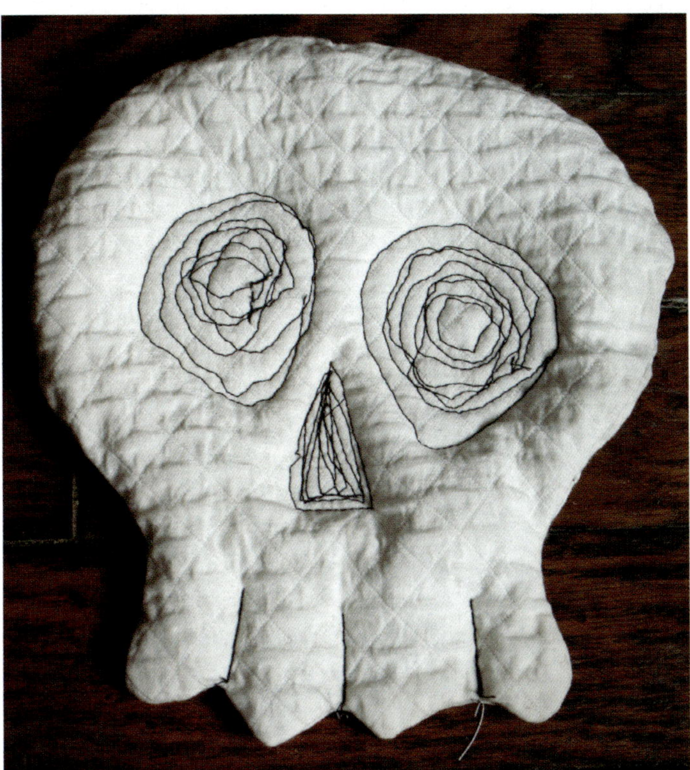

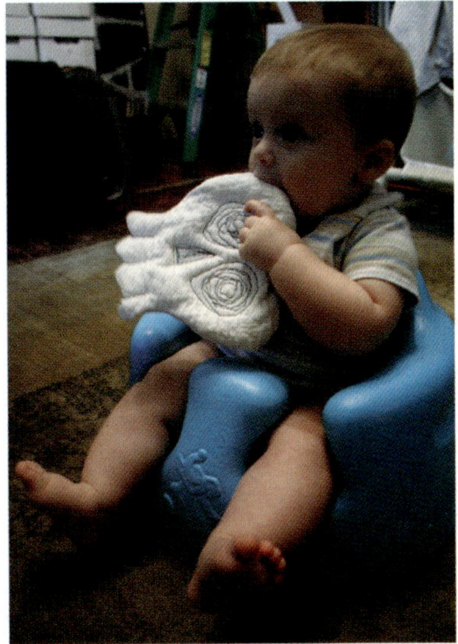

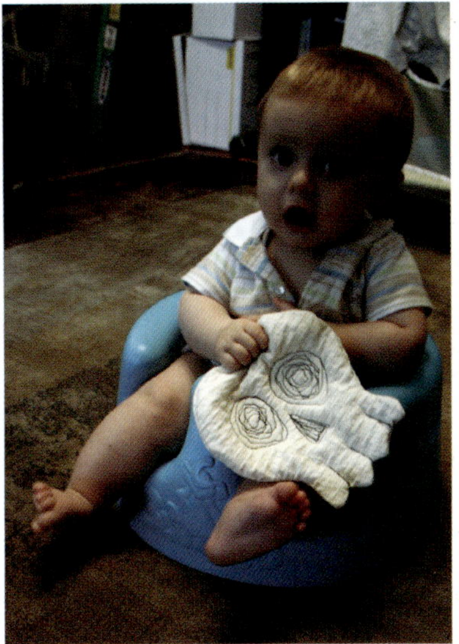

Baby Toy Skull Sewn Fabric Remnant (a.k.a. "Baby's First Death's Head", inspired by a Slowpoke cartoon by Jen Sorensen)

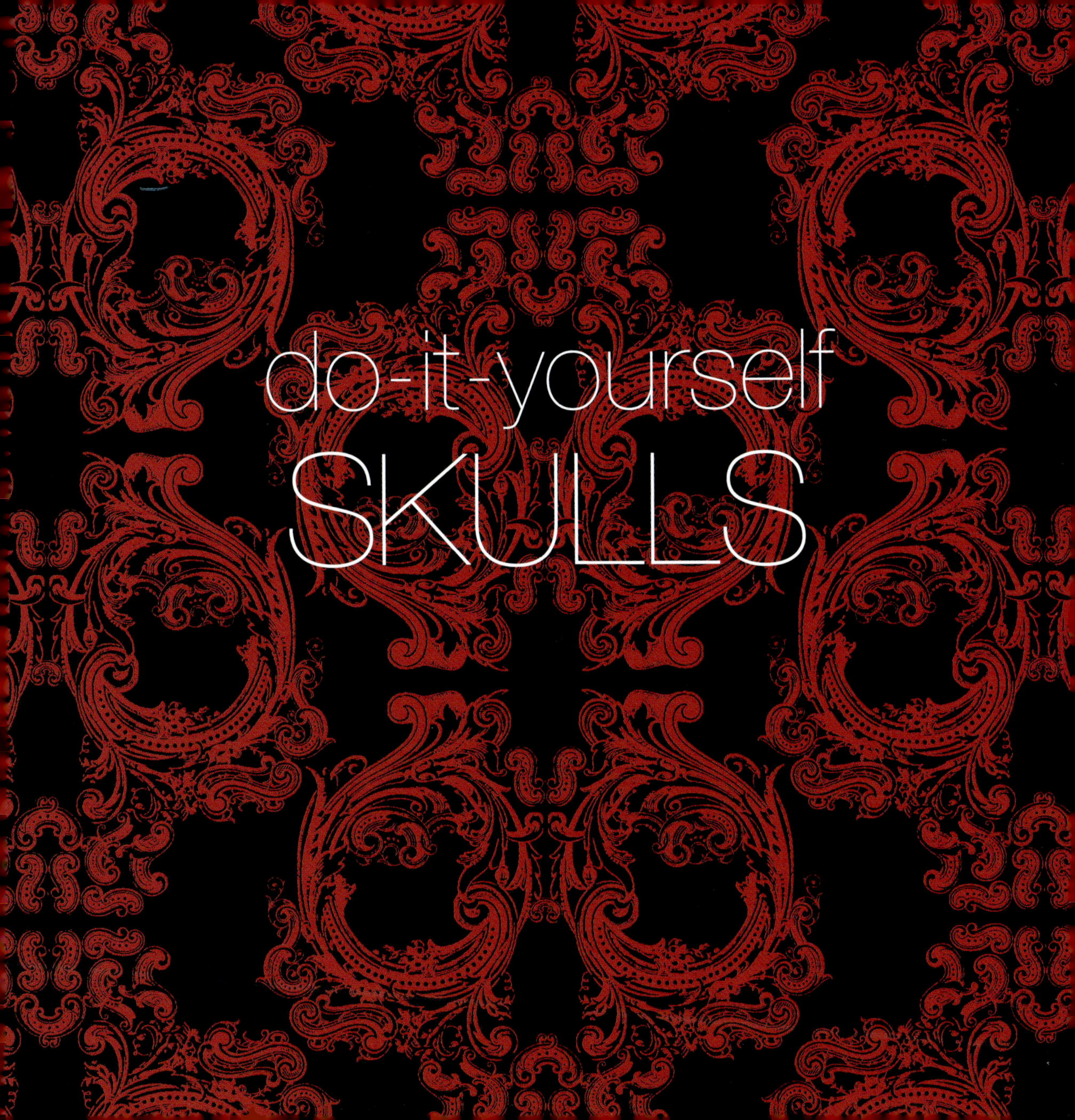
do-it-yourself
SKULLS

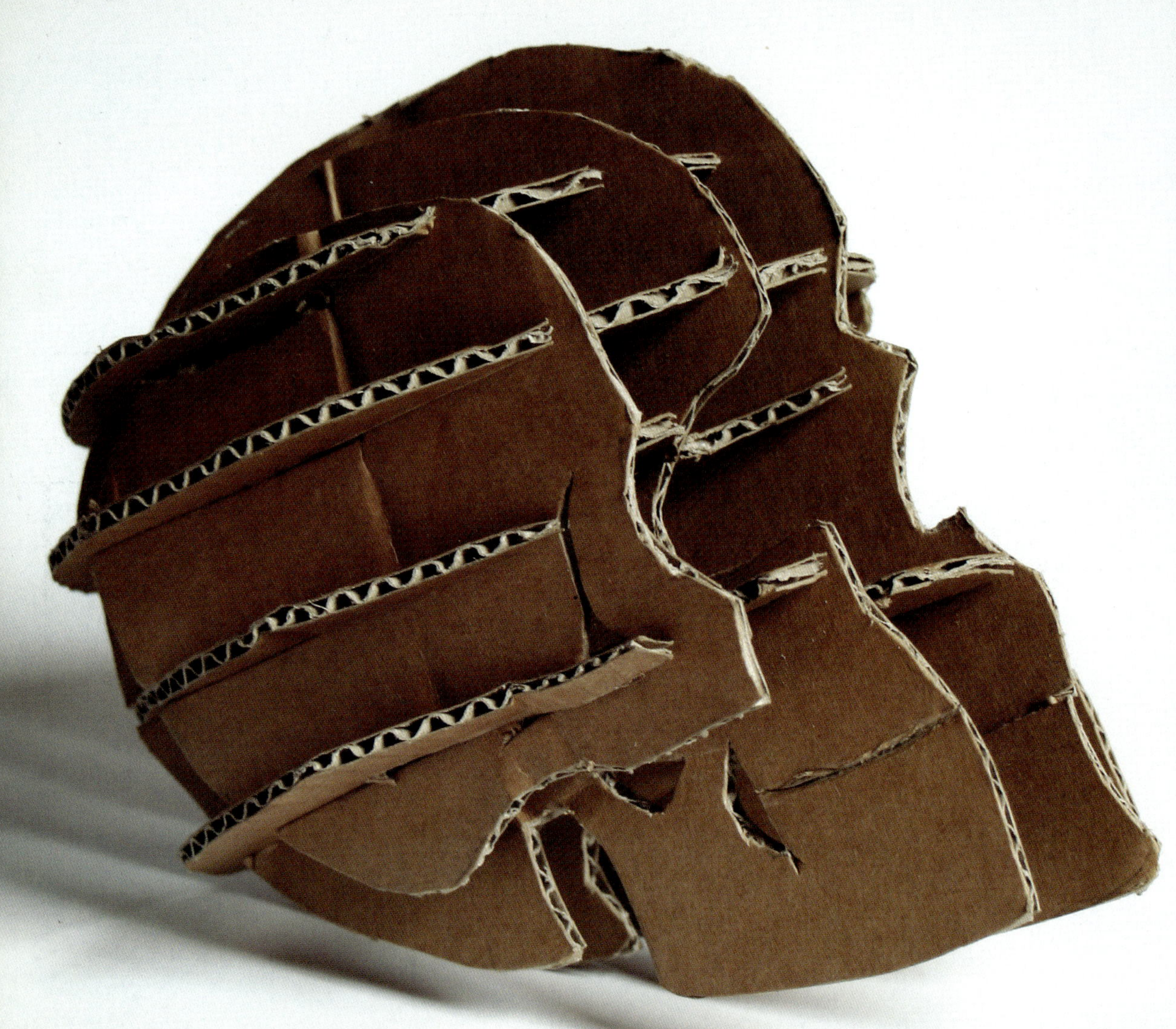

Make your own

Grid Skull

Make this skull any size by reproducing the pattern pieces larger or smaller. Be sure to adjust the width of the slots to match the thickness of the material you use.

YOU'LL NEED

- 9 pattern pieces
- Cardboard, poster board, or cardstock
- Sharp scissors or craft knife
- Craft glue

1 Enlarge the pattern pieces.

2 Glue them to your choice of heavy paper, and then cut out all the components.

3 Cut slots as shown. Width should match thickness of material used.

4 Slide horizontal pieces onto center piece, then add vertical pieces, moving from the center outward.

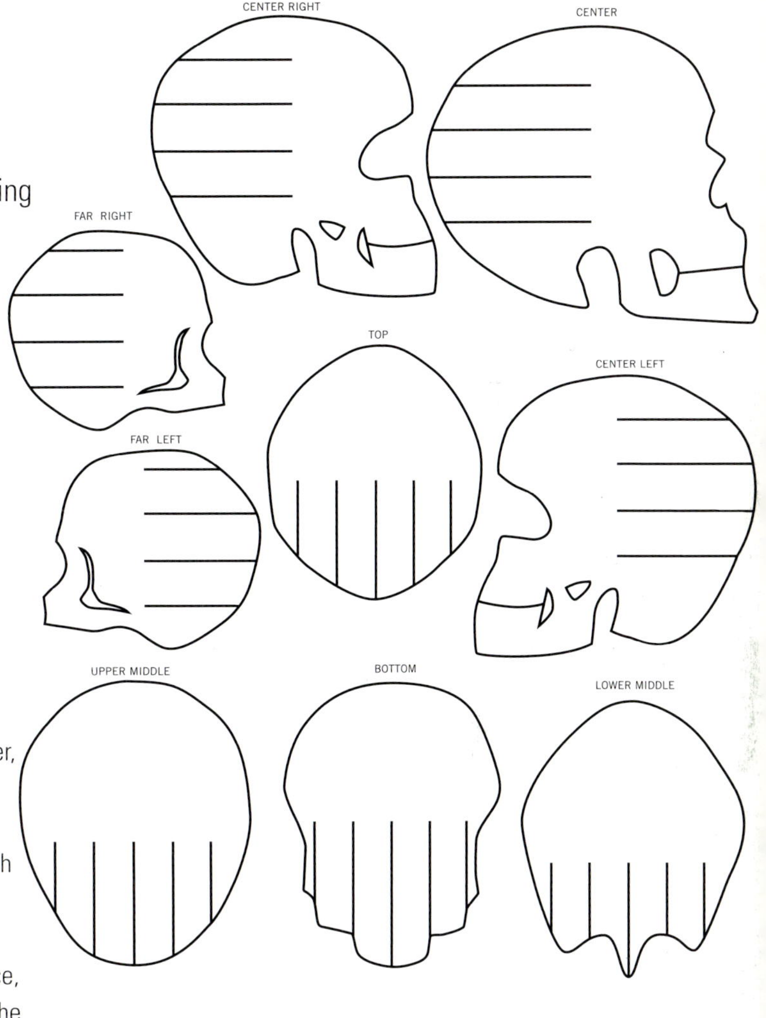

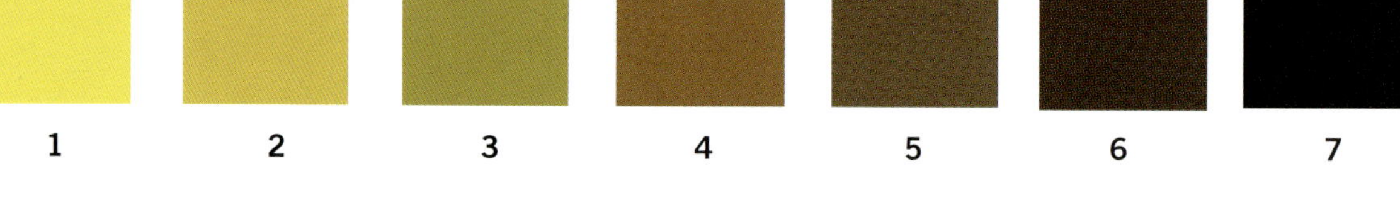

Paint your own Skull-by-Numbers

You may be surprised by the resulting image when you're done painting this skull. Feel free to experiment with different color palettes; just be sure to keep the value (lightness and darkness) similar.

YOU'LL NEED

- Pattern, enlarged as desired
- Colored markers, paint, crayons, etc.

1 Enlarge the pattern to the size you want.

2 Match the provided color swatches with your choice of painting materials, and follow the numbers!

A printable PDF of this pattern can be downloaded from www.skulladay.com.

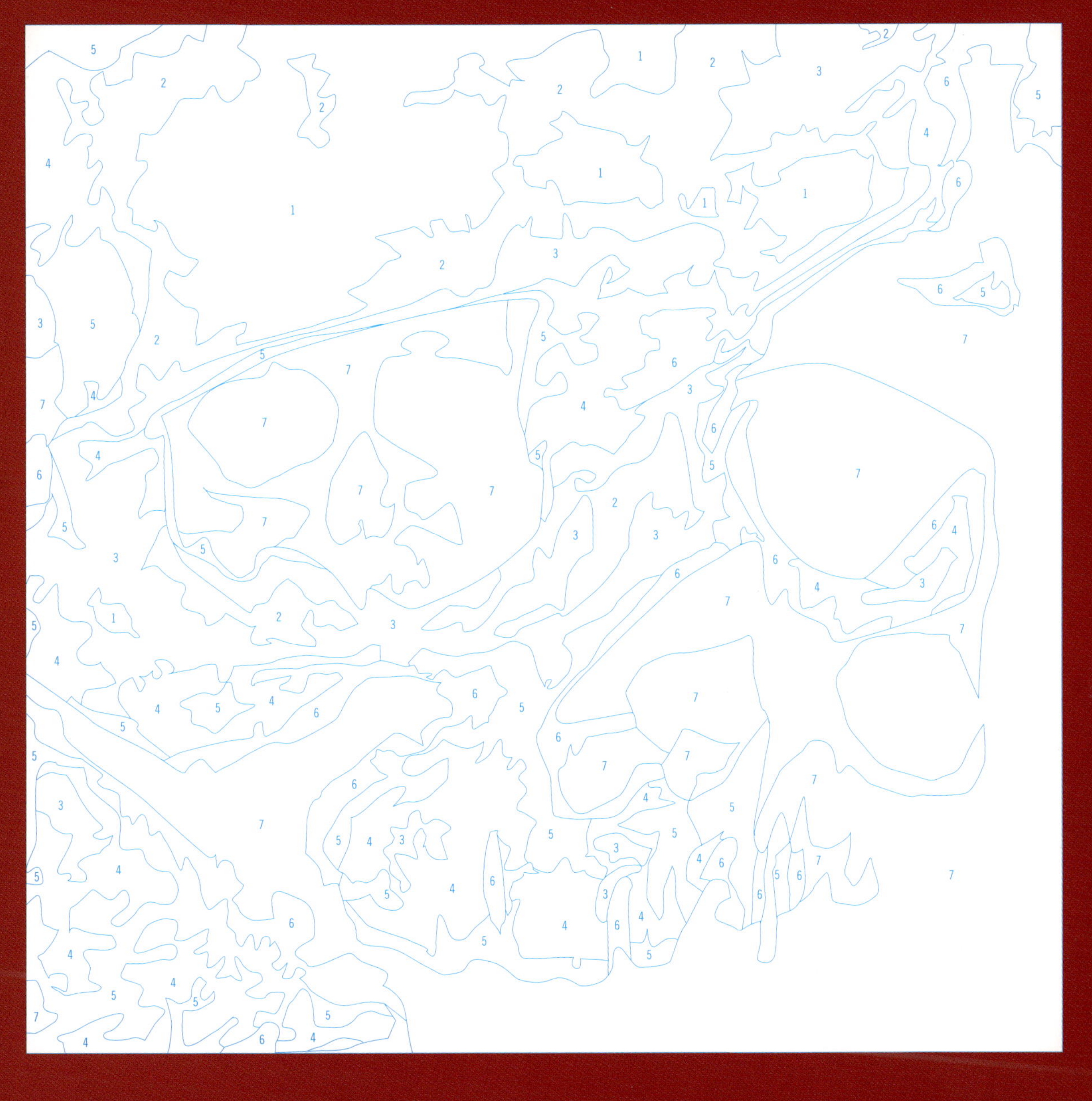

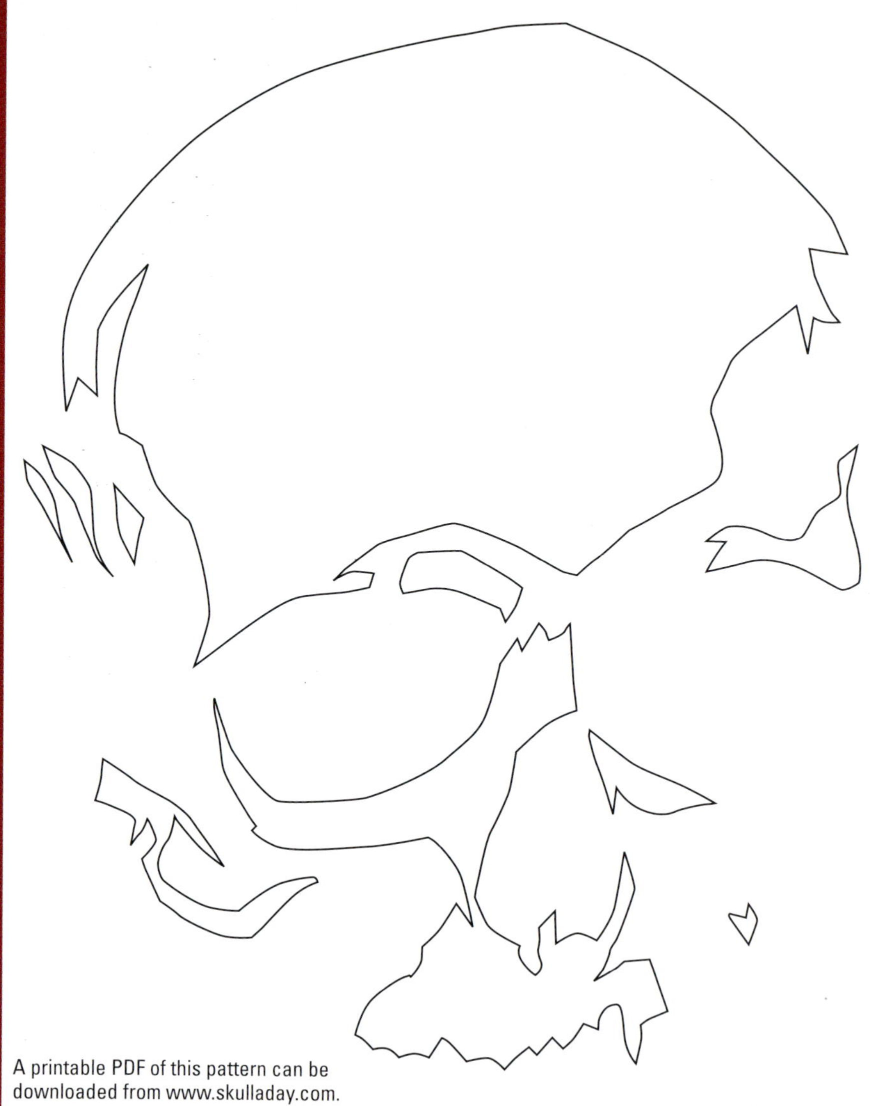

A printable PDF of this pattern can be downloaded from www.skulladay.com.

Make your mark with

Stencil Skull

Everyone needs a good skull stencil. Stencil the design onto cardstock, clothing, pillows, T-shirts, walls—whatever you want. The image looks best with a light color on a dark background.

YOU'LL NEED

- Stencil pattern
- Craft glue or easy-release tape
- Heavy paper or stencil plastic
- Craft knife
- Acrylic paints in your choice of colors
- Stencil brush

1 Copy and enlarge the stencil to the desired size.

2 Glue or tape it to a piece of heavy paper or stencil plastic.

3 Carefully cut out the pieces with the craft knife.

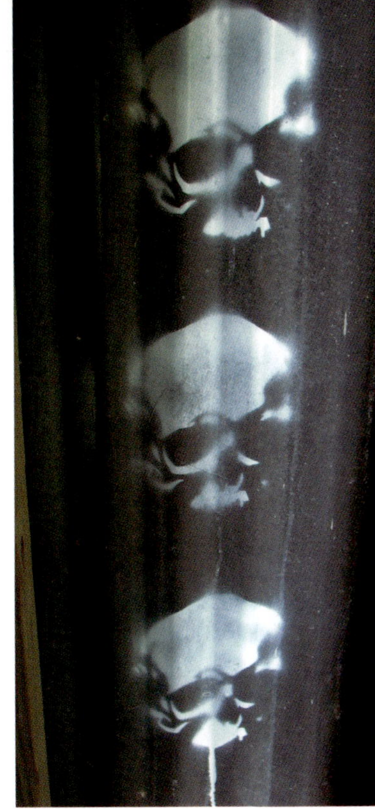

Spray paint on corrugated metal. (The stencil image is based on a photo I took in the Catacombs of Paris.)

4 Use a small amount of paint (the brush should be almost dry to the touch), and stencil the image onto the surface of your choice. If you've never used paint and a stencil before, it's a good idea to practice on a scrap of paper or fabric. This stencil can also be used with bleach on fabric, or powdered sugar on cakes!

Make your own
Paper Skull

You can use this pattern to make skulls with your own designs on them.

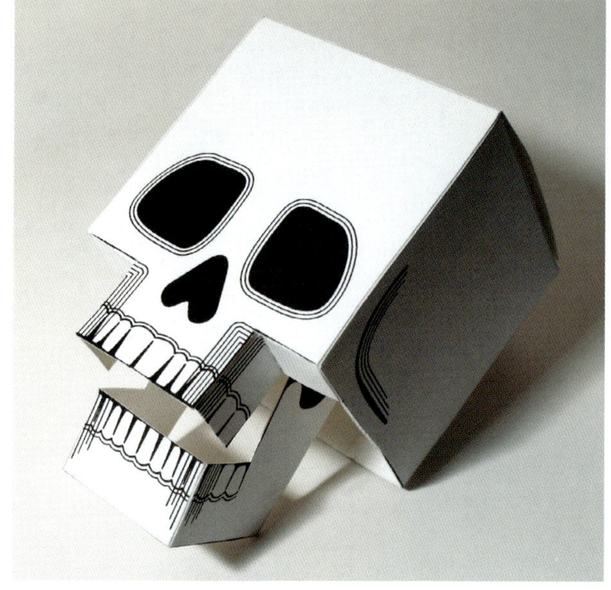

YOU'LL NEED

- Pattern pieces, copied and enlarged onto cardstock
- Scissors
- Craft glue or double-stick tape

1 Cut, score, and fold the pattern pieces as indicated.

2 Cut out the areas marked in red.

3 Use a small amount of glue or tape on the areas with an X.

4 Fold each piece individually, and leave the top of the skull open.

5 Slide the spindle through the jaw holes, and place the jaw into the bottom of the skull, with the lever coming out the slot in the back. Slide the spindle through the triangular holes in the tabs inside the skull (see the diagram).

6 Close the top of the skull. Move the lever in the back to open and close the jaw.

A printable PDF of this pattern can be downloaded from www.skulladay.com.

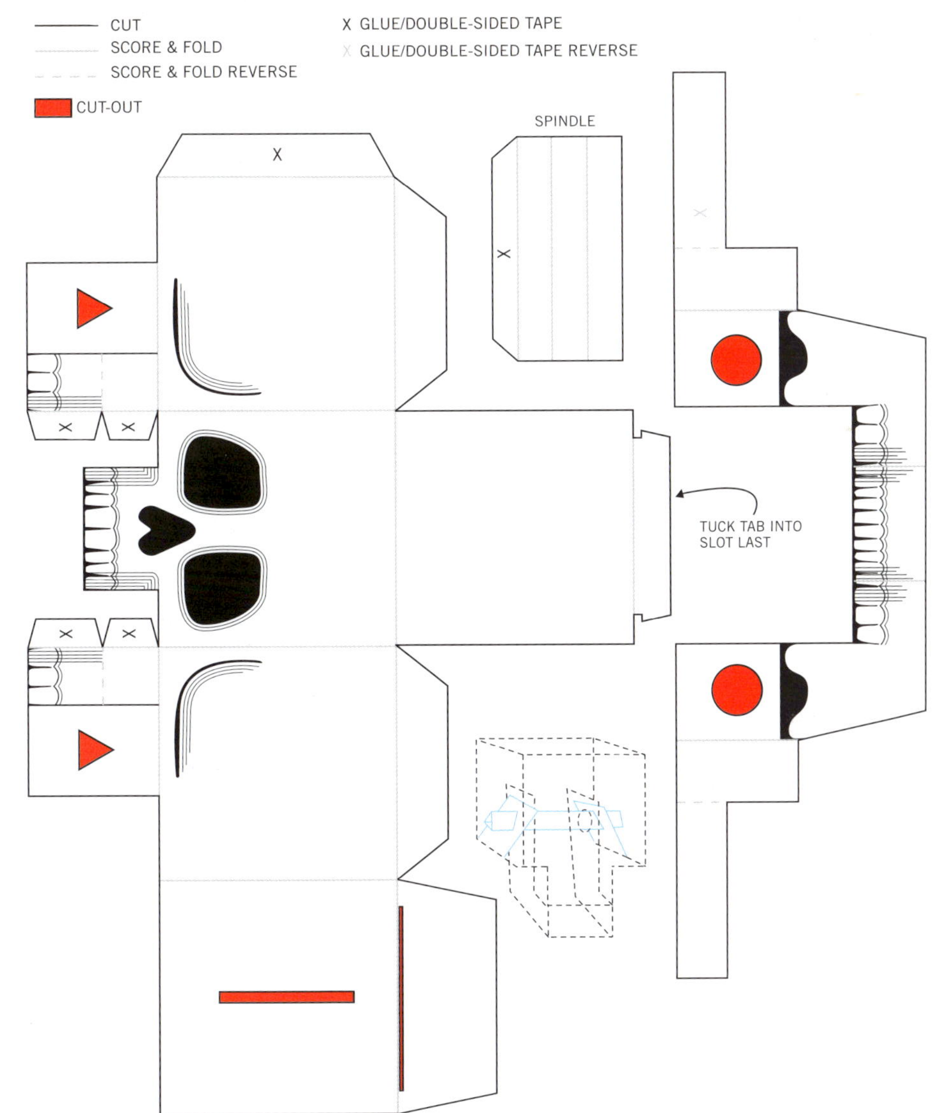
CUT
SCORE & FOLD
SCORE & FOLD REVERSE
X GLUE/DOUBLE-SIDED TAPE
X GLUE/DOUBLE-SIDED TAPE REVERSE
CUT-OUT
SPINDLE
TUCK TAB INTO SLOT LAST

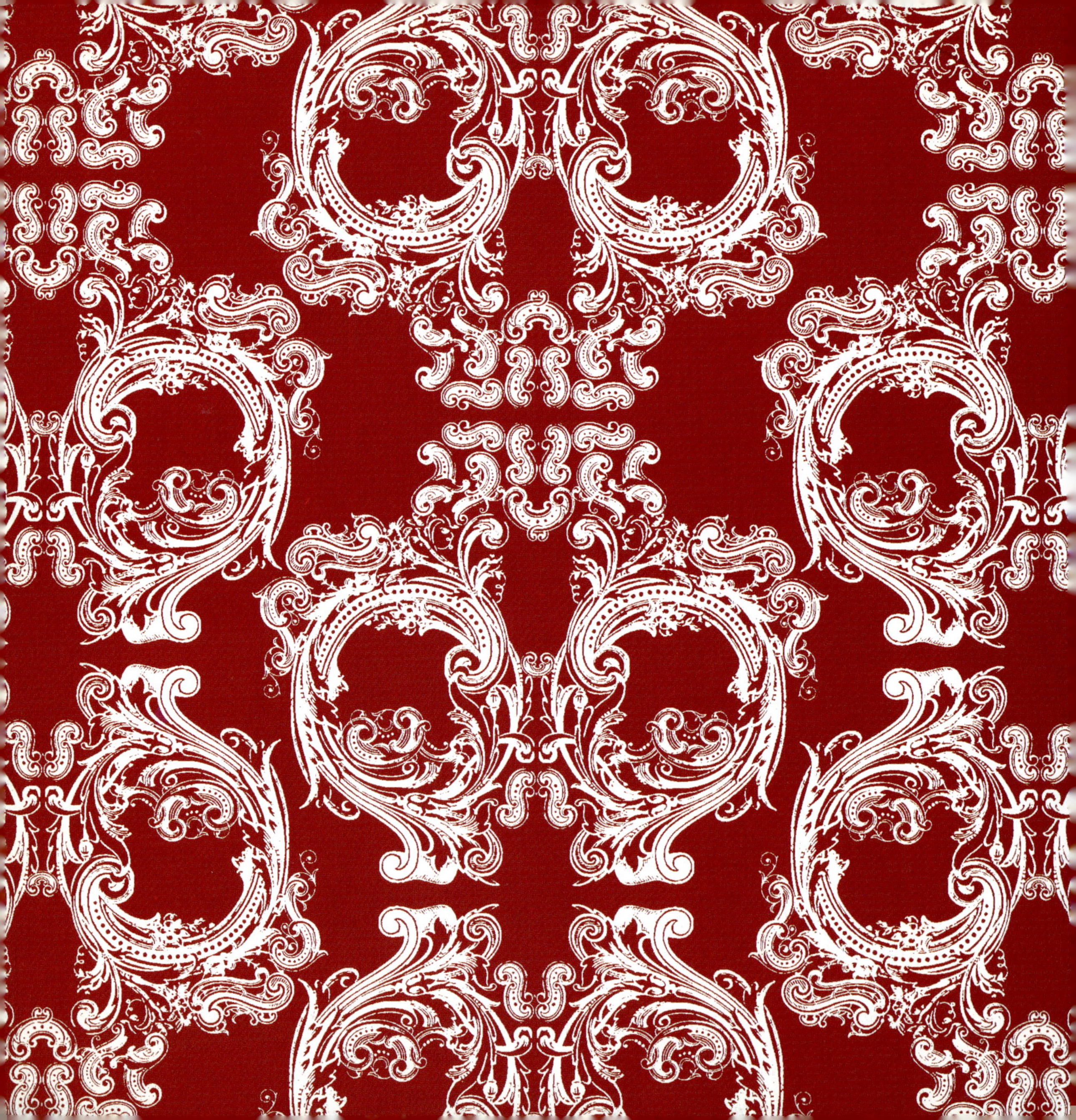

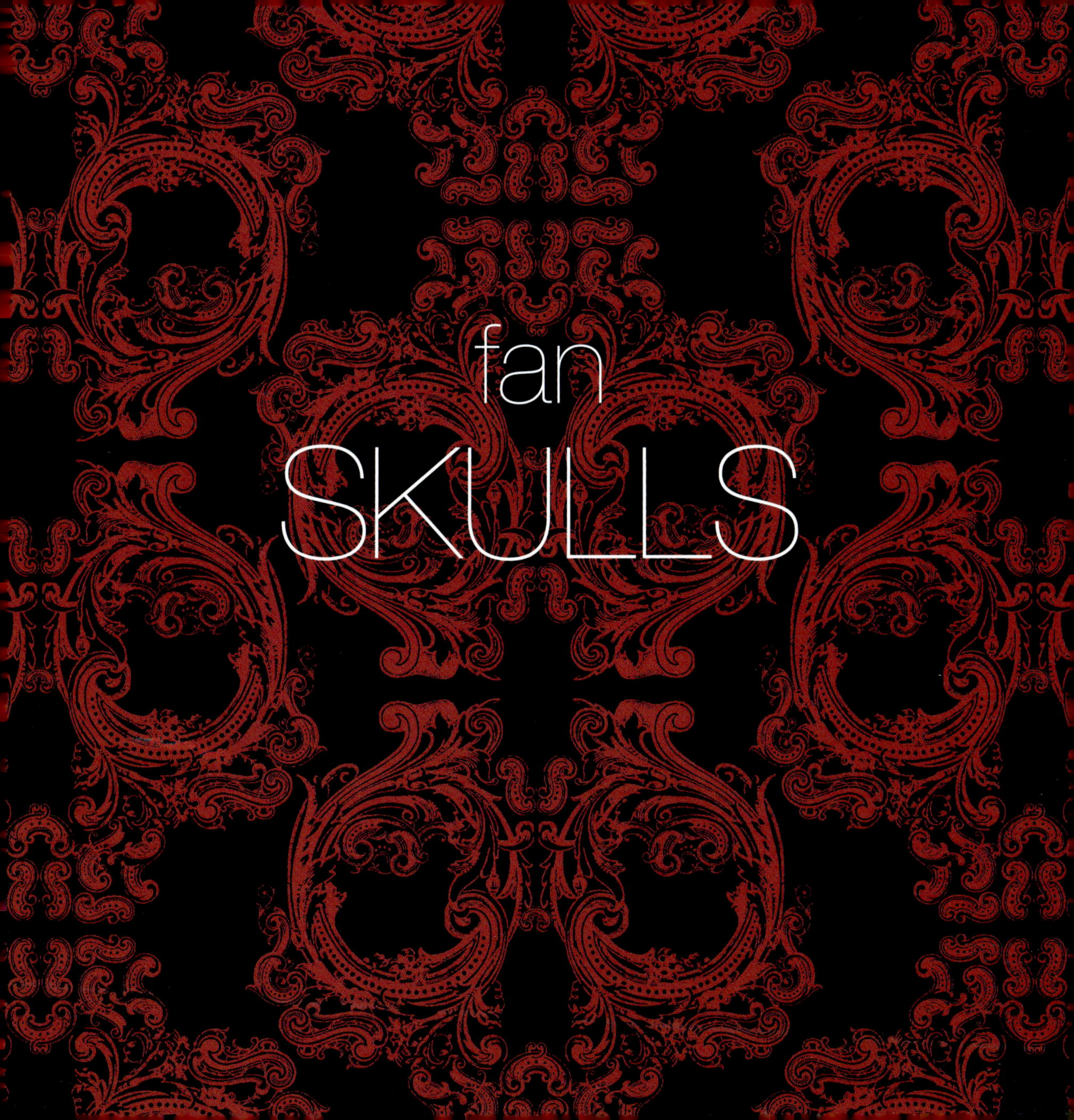
fan
SKULLS

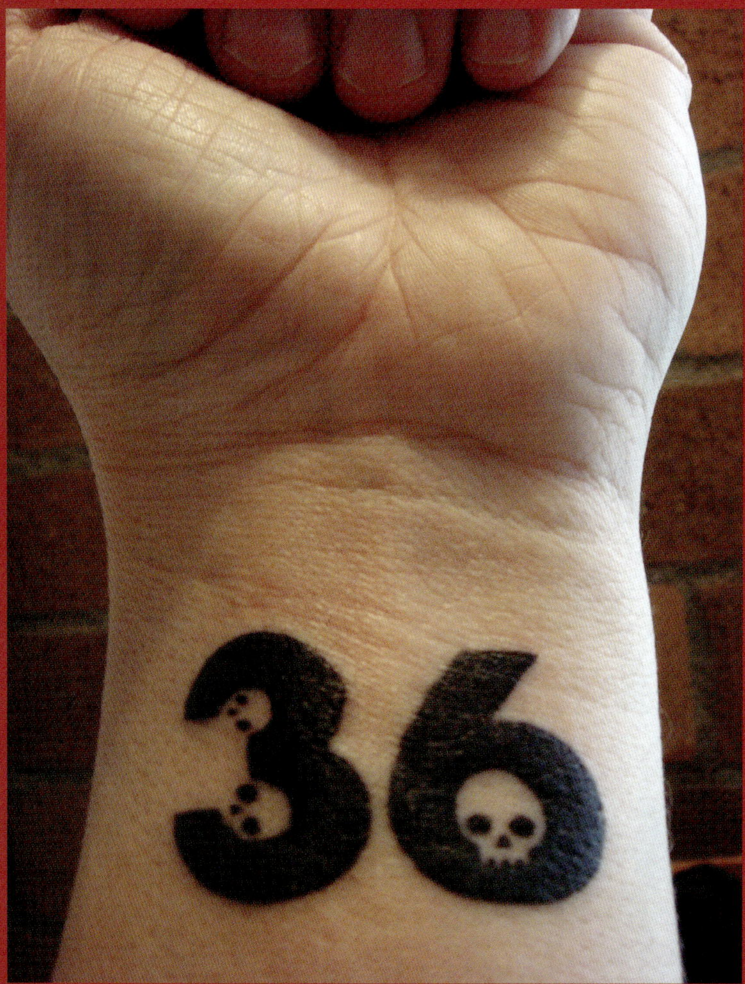

A.G. Wilson in Seattle got a tattoo based on my Skullphabet font!

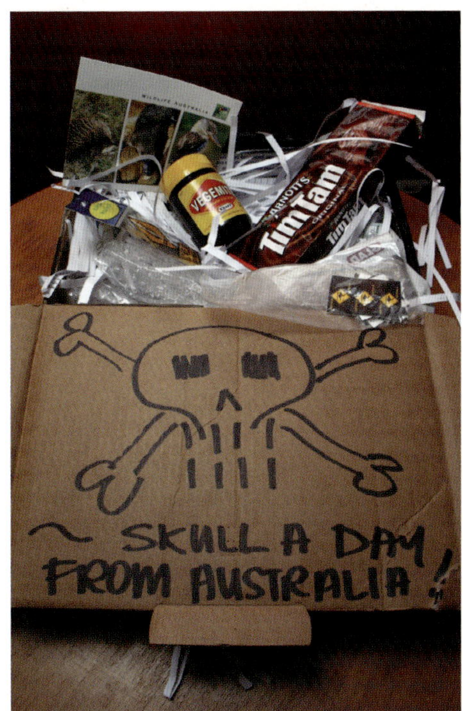

A box of goodies came from Beck in Australia. Of course, I used the Vegemite to make a skull.

Australian Kim Upton made this hat and matching mittens, which incorporate my 8 x 8 skull, as a gift for me.

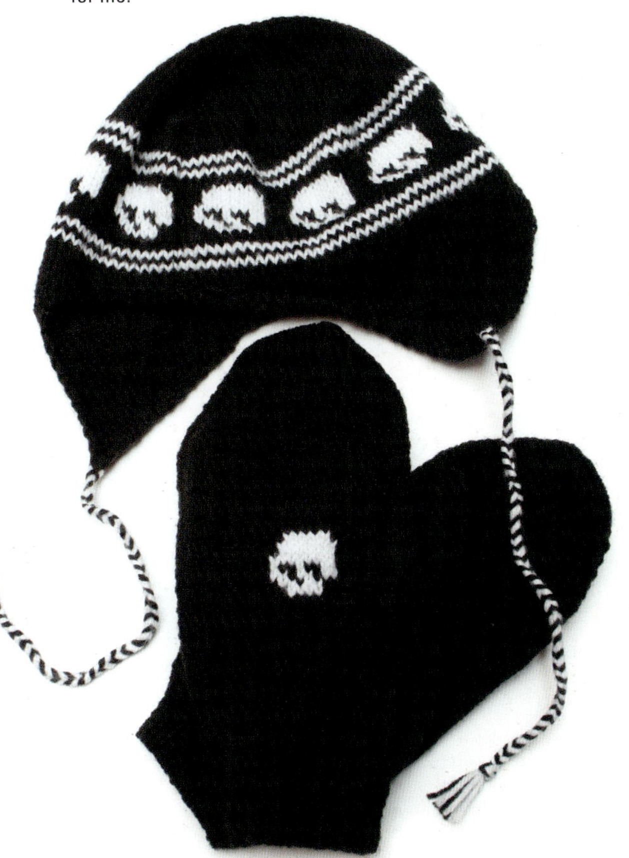

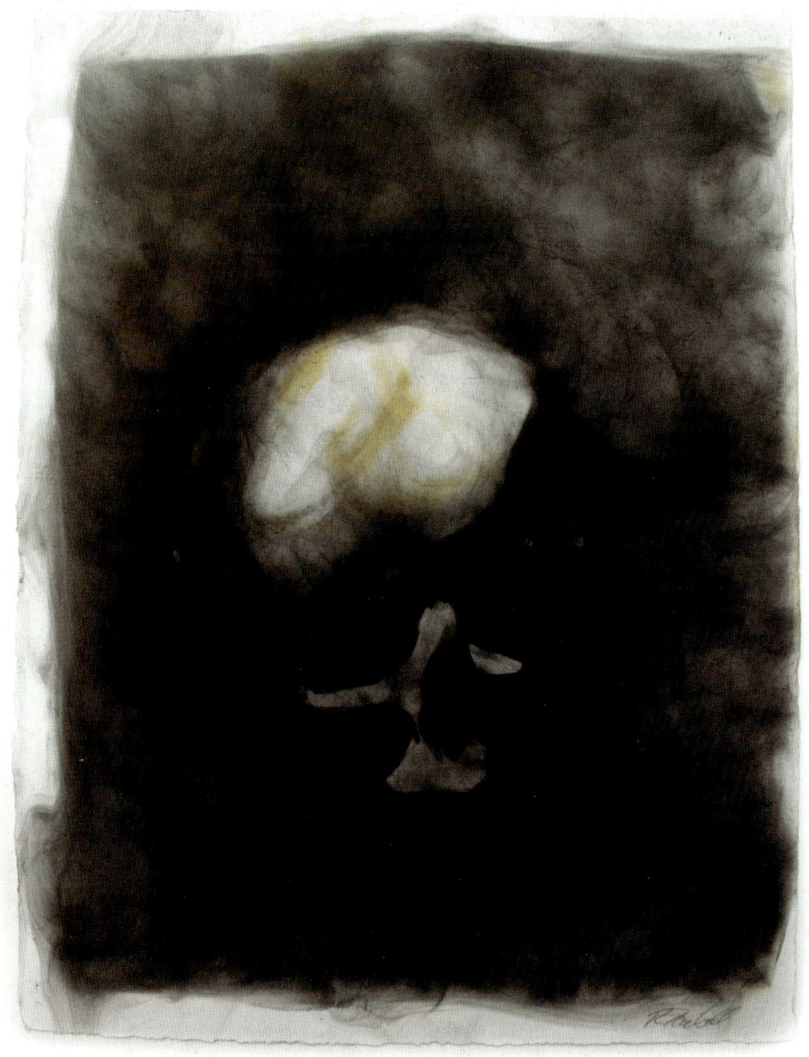

Virginia artist Rob Tarbell made this smoke art for me using one of my stencils.

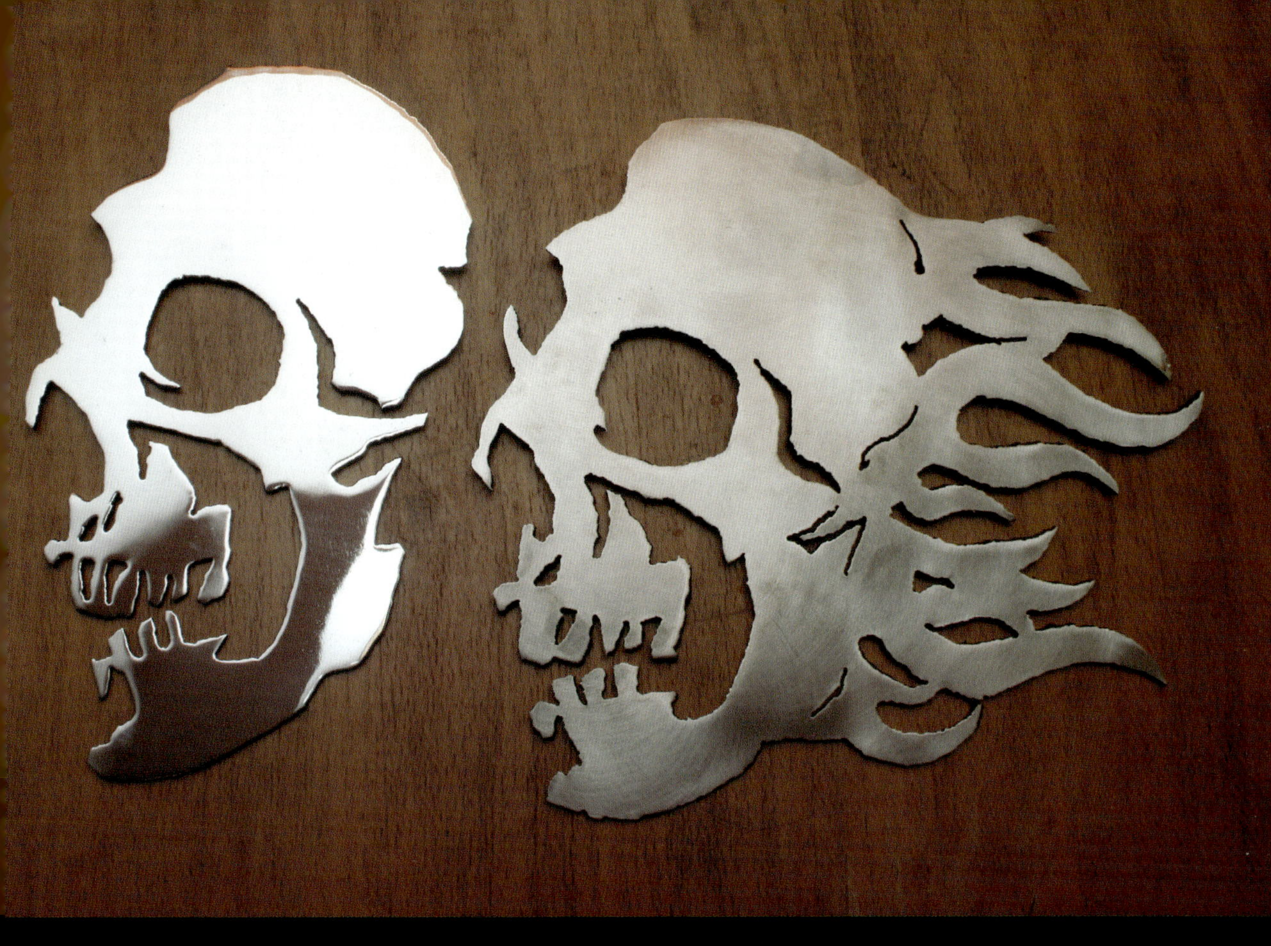

Artist Larry Pearson in Iowa mailed me these beautiful steel pieces that he created based on one of my stencils!

By Oliver Larsen

12-year-old Oliver Larson in Denmark made this typewritten skull based on one of my stencils!

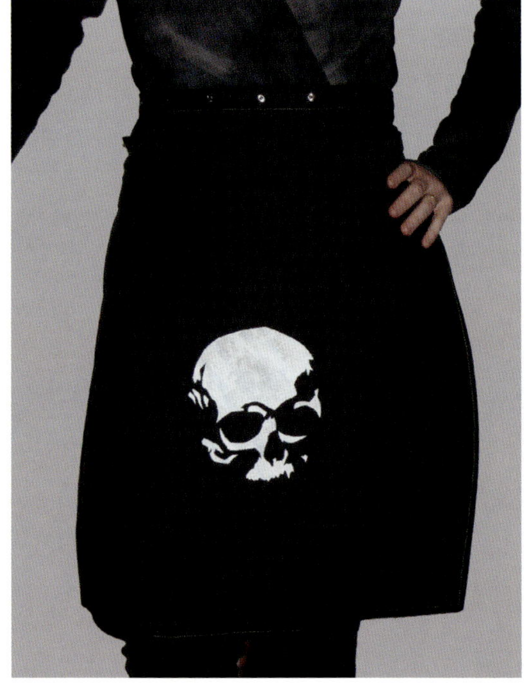

Patricia Dennis in Buffalo, NY, made this lovely apron using one of my stencils.

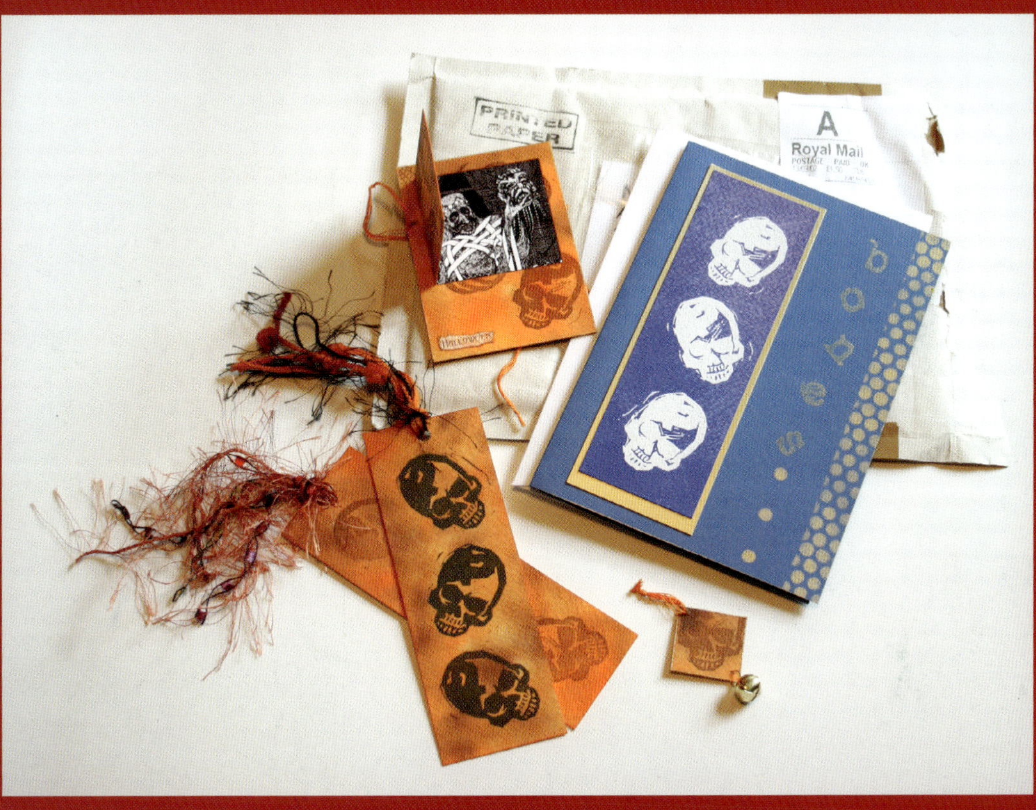

Gillian McMurray in Scotland sent me these beautiful paper goods that she made with a custom rubber stamp based on one of my skulls!

Acknowledgments

Creating a daily art project (and then turning it into a book) is no easy task. Doing that while trying to run a full-time business and still find time to eat meals would have been impossible were it not for the support of my amazing friends and family. I'm so grateful to have you all be a part of this ridiculous journey of mine.

A hearty thanks to everyone who helped along the way: Sara Heifetz and Carlos Collazo, Marilyn Scalin and Roger Davis, Chelsea Kostek, Cheryl Silver, Kristin Murray, Kit French, Charlie Bonét and Tere Hernández-Bonét, Christiana LaMountain & Russ Reed, Carra Rose and Spencer Hansen, Amy McFadden, Leah Engel, Anna von Gehr and Rob Tarbell, Jude Schlotzhauer, Lisa Taranto, Matt Deans, Shelia Gray and Philip Perrine, Nathan Wender, Adrien and Kenny Hamilton, Chris Boarts-Larson and Stig, Christine Colby, Betsy/Betty Migliaccio and Dave Burley, Mimi Regelson and Tsondu Kikhangparra, Sarah Luddy and Peter Fiala, Zeev Krieger, Christopher Humes, Eliza Skinner, Traci and Chris Whitley, Julien and Ekin Aksoy, Sherry and John Petersik, Kris Iden and Christian Wiendl, Vanessa Bertozzi and all the fine folks at the Etsy Labs, Teddy Blanks and Ross Harman A.K.A. The Gaskets, Patrick Godfrey and Velocity Comics, Clark Whittington and all of the awesome Art-o-mat artists, Cory Doctorow and Xeni Jardin of Boing Boing, Mark Conahan for his invaluable work translating the Skullphabet into a functional font, and of course all the folks who dressed up for my skull-themed Halloween party! My apologies to anyone I've forgotten, chalk it up to my addled brain and not to my lack of appreciation—hopefully I thanked you many times in person!

A special thanks goes out to the über-fans of Skull-A-Day, who made sharing my work with the world so incredibly rewarding: Kim Upton, Larry Pierson, Beck AKA Kitty, Joseph Biondi AKA b13, Monster Maniac, Justin Lovorn AKA Tatman, Dan Springer, Christiane Brandmaier, and most of all to the mysterious Citizen Agent!

Eternal gratitude belongs to Deborah Morgenthal, Susan McBride, and all the kind folks at Lark Books for the love and attention they've given my work.

A massive thanks, with cherries on top, to my fantastic agent Kate McKean, for asking me if I wanted to make a book.

My sister Mica is the epitome of awesome and none of my various and sundry projects including this one would be complete without her participation.

There are not enough thanks in the world for Jessica, who had a daily firsthand experience of the making of this project and never wavered in her support. I would most likely have lost my mind (entirely) if she hadn't been here for me.

I had the privilege of being raised in a house with a room dedicated to making art, and I can honestly say this project would not have been possible with out my wonderful parents Chuck and Mim.

Biography

Noah Scalin is an artist/activist living in Richmond, Virginia. When he is not making skulls, he runs Another Limited Rebellion, the internationally recognized, socially conscious design and consulting firm he founded. Noah also created the community supported agriculture group Sprout, and teaches Design Rebels, a course on socially conscious design, in the graphic design department at Virginia Commonwealth University.

Photo by Mark Mitchell